DESCARTES AND THE LAST SCHOLASTICS

DESCARTES
AND THE
LAST SCHOLASTICS

Roger Ariew

Cornell University Press

Ithaca and London

Copyright © 1999 by Cornell University

All rights reserved. Except for brief quotations in a review, this book, or parts thereof, must not be reproduced in any form without permission in writing from the publisher. For information, address Cornell University Press, Sage House, 512 East State Street, Ithaca, New York 14850.

First published 1999 by Cornell University Press

Library of Congress Cataloging-in-Publication Data

Ariew, Roger.
    Descartes and the last scholastics / Roger Ariew.
    p. cm.
    Includes bibliographical references and index.
    ISBN 0-8014-3603-6 (alk. paper)
    1. Descartes, René, 1596–1650. 2. Scholasticism. I. Title.
B1875.A65 1998
194—dc21   99-11953

Printed in the United States of America

Cornell University Press strives to use environmentally responsible suppliers and materials to the fullest extent possible in the publishing of its books. Such materials include vegetable-based, low-VOC inks, and acid-free papers that are recycled, totally chlorine-free, or partly composed of nonwood fibers. Books that bear the logo of the FSC (Forest Stewardship Council) use paper taken from forests that have been inspected and certified as meeting the highest standards for environmental and social responsibility. For further information, visit our website at www.cornellpress.cornell.edu.

Cloth printing   10 9 8 7 6 5 4 3 2 1

For MG and DG

Contents

Acknowledgments ix

List of Abbreviations xi

Introduction 1

1. Descartes among the Scholastics 7

PART I Context

2. Descartes and the Scotists 39

3. Ideas, in and before Descartes 58
   with Marjorie Grene

4. The Cartesian Destiny of Form and Matter 77
   with Marjorie Grene

5. Scholastics and the New Astronomy on the Substance
   of the Heavens 97

PART II Debate and Reception

6. Descartes, Basso, and Toletus: Three Kinds of Corpuscularians 123

7. Descartes and the Jesuits of La Flèche: The Eucharist 140

8. Condemnations of Cartesianism: The Extension and Unity
   of the Universe 155

9. Cartesians, Gassendists, and Censorship	172
10. Scholastic Critics of Descartes: The *Cogito*	188
Appendix: Gilson's *Index* Indexed	207
Bibliography	211
Index	223

Acknowledgments

This research was assisted by grants and fellowships from the National Endowment for the Humanities, an independent federal agency, the National Science Foundation (grant no. DIR-9011998), and the Foundation for Intellectual History. None of it would have been possible without the funds these marvelous institutions made available—nor would I have been able to afford the research time at the Bibliothèque Nationale in Paris. I am also grateful to the Bibliothèque for allowing me to use its wonderful facilities and for providing me with microfiches of countless volumes. Moreover, I thank the Department of Philosophy and the College of Arts and Sciences at Virginia Polytechnic Institute and State University for releasing me from some of my teaching obligations and for making available an ad hoc research grant to purchase reproductions of necessary rare documents. (In particular, I thank Robert Bates, Dean, and Adelene Kirby, Research Coordinator for the College, for their efforts on behalf of this project.)

Portions of this book were read (at different stages of their composition) at many universities, workshops, and conferences. I am indebted to all these audiences for their questions and comments. I am also grateful to numerous publishers for allowing me to use some of my previously printed materials. An earlier version of Chapter 1 was published as "Descartes and Scholasticism: the Intellectual Background to Descartes' Thought," in *Cambridge Companion to Descartes*, ed. John Cottingham (Cambridge: Cambridge University Press, 1992), pp. 58–90. Chapter 2 is being issued as "Descartes and the Scotists," in *The Connaught Descartes*, ed. Brian Baigrie, André Gombay, and Calvin Normore (Toronto University Press). Chapter 3 was published as "Ideas, in and before Descartes," *Journal of the History of Ideas* 56 (1995): 87–106, and Chapter 4 as "The Cartesian Destiny of Form and Matter," *Early Science and Medicine* 3 (1997): 300–325, both coauthored by Marjorie Grene. A very early version of Chapter 5 appeared as "Theory of Comets at Paris during the Seventeenth Century," *Journal of the History of Ideas* 53 (1992): 355–72. Chapter 6 is forthcoming as "Descartes, Basson et la scolastique renaissante," in *Descartes et la Renaissance,* ed. Emmanuel Faye, and Chapter 7 as "Les premières tentatives vers une scolastique cartésienne: La correspondance de Descartes et les Jésuites de La Flèche sur l'Eucharistie," in *Momenti della biografia intellettuale di Descartes nella Correspondance,* ed. Jean-Robert Armogathe and Giulia Belgioioso. A portion of Chapter 8 was published as "Les *Principia* en France et les condamnations du cartésianisme," in *Descartes: Principia Philosophiae (1644–1994),* ed. Jean-Robert Armogathe and Giulia Belgioioso (Naples: Vivarium, 1996), pp. 625–40. A version of Chapter 9 was issued as "Damned If You Do: Cartesians and Censorship, 1663–1706," *Perspectives on Science* (1994): 255–74 (copyright © 1994 by the University of Chicago), and of Chapter 10 as "Critiques scolastiques de Descartes: Le *cogito,*" *Laval Théologique et Philosophique* 53, no. 3 (1997): 587–604.

Moreover, this work would not have been feasible without the excellent conversation and good advice of many colleagues, graduate students, and colleagues-at-large (I hope that those I may have forgotten will forgive my oversight): Jean-Robert Armogathe, Peter Barker, Giulia Belgioioso, Jean-Marie and Michelle Beyssade, Constance Blackwell, Ann Blair, Lawrence Brockliss, Frederick de Buzon, Vincent Carraud, John Cottingham, Edwin Curley, Dennis Des Chene, Mordechai Feingold, Alan Gabbey, Matthew Hettche, Eric Lewis, Jean-Luc Marion, John Murdoch, Steve Nadler, Joseph Pitt, Tad Schmaltz, Theo Verbeek, Eric Watkins, and Robert Westman.

Marjorie Grene has read the entire manuscript and made numerous comments requiring both substantive and stylistic revisions. Susan Andriette Ariew has also read all my prose. The two of them have always been my most severe critics—something for which I am very thankful.

The volume owes a particular debt to two scholars and friends: the previously mentioned Marjorie Grene, with whom I have written a couple of the essays and discussed the history of philosophy daily for almost a decade, and Daniel Garber, with whom I have collaborated on the investigation of seventeenth-century philosophy for an even longer period. (I look forward to the continued fruits of our labors.) It has gotten to be so that I no longer know whether something I say was originally mine or theirs, so close are our views, so much have they influenced my thinking. Thus, I think it is fitting that this book be dedicated to them, given that it is theirs already.

R. A.

Blacksburg, Virginia

Abbreviations

Works of Descartes have been identified in the notes by the following abbreviations. Unless otherwise indicated, translations are the author's.

ACS Ariew, Roger, John Cottingham, and Tom Sorell, eds. and trans. 1998. *Descartes' Meditations: Background Source Materials.* Cambridge, Cambridge University Press

AT Descartes, René. 1964–74. *Oeuvres de Descartes.* Ed. C. Adam and P. Tannery. 2d ed. Paris, Vrin.

CSM Descartes, René. 1984–85. *The Philosophical Writings of Descartes.* Vols. 1 and 2. Trans. J. Cottingham, R. Stoothoff, and D. Murdoch. Cambridge: Cambridge University Press.

CSMK Descartes, René. 1991. *The Philosophical Writings of Descartes.* Vol. 3. Trans. J. Cottingham, R. Stoothoff, D. Murdoch, and A. Kenny. Cambridge: Cambridge University Press.

DESCARTES AND THE LAST SCHOLASTICS

Introduction

A philosophical system cannot be studied adequately apart from the intellectual context in which it is situated. Philosophers do not usually utter propositions in a vacuum, but accept, modify, or reject doctrines whose meaning and significance are given in a particular culture. Thus, Cartesian philosophy should be regarded, as indeed it was in Descartes's own day, as a reaction against, as well as an indebtedness to, the scholastic philosophy that still dominated the intellectual climate in early seventeenth-century Europe. But it is not sufficient, when discussing Descartes's relations with Scholastics, simply to enumerate and compare the various Cartesian and scholastic doctrines. To understand what set Descartes apart both from the Scholastics and from other innovators, one has to grasp the reasons behind the various opinions; but, beyond that, one also has to understand the intellectual milieu in which these reasons played a role, to see what tactical measures could have been used to advance one's views or to persuade others of them. This is the common theme linking the chapters that follow.

The theme is exemplified in Chapter 1, in which I contrast Descartes's attitude toward scholastic philosophy as revealed in his correspondence with the attitude in his published works. I then present enough background about Jesuit pedagogy and philosophy to begin to understand Descartes's attempt to gain favor among members of that order. I depict a few skirmishes between Descartes and Scholastics (including Jesuits) to capture the flavor of such exchanges. Perhaps the most interesting lesson that can be learned by looking at Descartes's relations with Scholastics is the sheer power and authority of Aristotelianism during the seventeenth century. This authority is manifest in the official reactions to Descartes's philosophy. The pattern of this first chapter is repeated throughout the book: I move from within Cartesian philosophy to its intellectual context in seventeenth-century France, then to the living philosophical debate waged by Descartes and his contemporaries, and finally to the first reception of Cartesian philosophy as another path (though an indirect one) one can take in order to understand Descartes's philosophy as it was originally intended. Because of this common pattern, the chapters are grouped into two parts, depending upon whether a given essay more particularly stresses contextual issues or live debate and reception.

These chapters are not offered as close examinations of Descartes's philosophy, they are merely intended as a first step toward such an examination. Again, before asking what philosophers hold and why, we need to familiarize ourselves with the philosophical options open to them and the language used to express such options. We need to understand the meaning those terms had in that particular culture, the significance of various philosophical views for the culture, and so on. We cannot simply assume that these things are the same for any culture as they are for us. Thus, in the search for clues to an adequate understanding of any particular philosophy, we are bound to investigate both its context—in its social and intellectual dimensions—and its immediate reception. Despite the vast number of essays presented as close examinations of Descartes's philosophy, that task, I suggest, has not been sufficiently accomplished.

Part I, "Context," exhibits the differences and similarities between the doctrines of Descartes, the Jesuits, and other Scholastics in seventeenth-century France. I demonstrate in Chapter 2 that the philosophical context in France during the early 1600s was predominantly Scotist and not Thomist. This fact has been obscured in part because Etienne Gilson, the great French Cartesian commentator, wrote as if all seventeenth-century textbook authors were Thomists: Gilson's *Index scolastico-cartésien* compared Descartes with Thomas Aquinas, the Coimbrans, Francisco Suarez, Franciscus Toletus, Antonius Rubius, and Eustachius a Sancto Paulo—that is, Gilson compared Descartes with Thomas, (Iberian and Roman) Jesuits, and Eustachius, a Paris doctor. Contra Gilson, an analysis of Eustachius's works quickly shows that every doctrine one would call Scotist was held by him: the univocity of being; matter having being apart from form; space as radically relational; time as independent of motion; the plurality of forms; the theory of distinctions, including the formal distinction; individuation as *haecceity*, that is, a form; being in general as the proper object of the human intellect; and so forth. It is clear that Eustachius was propounding common Parisian doctrines (with others, such as Charles François d'Abra de Raconis and Scipion Dupleix), that these opinions became dominant (even with later Jesuits such as Pierre Gaultruche), and that they were often issued self-consciously as anti-Thomist—thus, it is also evident that the categories "Thomist" and "Scotist" were actors' categories for seventeenth-century Scholastics. In this chapter, finally, I suggest ways in which this knowledge might open up interpretive paths for understanding Descartes himself.

In Chapter 3, written with Marjorie Grene, I discuss Descartes's concept of idea and the scholastic context from which it arose. I examine the use of the term in the writings of four seventeenth-century philosophers—Eustachius, Jean Crassot, de Raconis, and Rudolph Goclenius—and contrast it with its traditional use in the philosophical corpus of seventeenth-century Scholasticism (in the *Corps de toute la philosophie* of Théophraste Bouju, for example). I conclude that Descartes drew on the current seventeenth-century literary and philosophical usages as represented by Eustachius, Goclenius, and de Raconis; that by calling on the ideas in God's mind as his source, Descartes set ideas free from their connection to sensation; and that there is precedent in the philosophical literature for Descartes's insistence on the truth of ideas.

Chapter 4, also written with Marjorie Grene, is an examination of the allegedly enormous differences between strict hylomorphism and Cartesianism on form and matter. For a strict hylomorphist, matter and form cannot be separated, but for a Cartesian the two are really distinct. For a strict hylomorphist, similarly, form is the principle of being and matter the principle of individuation, but for a Cartesian, the mind—a form—is the principle of individuation for persons, if anything is. However, the break between Cartesianism and Scholasticism is not as severe as these differences would indicate, if seventeenth-century Scholasticism is taken into account. For many reasons, the late Aristotelians also broke with Aristotle, accepted the reality of matter without form and form without matter, and understood form as the principle of individuation. In addition, the intellectual landscape of seventeenth-century philosophy was not limited to the properly scholastic; there were anti-Aristotelian options (some corpuscularian, others not) available before Descartes. Given that the gulf between the Schoolmen and *novatores* like Descartes was not so wide, the way was open for certain compromises: a variety of scholastic restatements of Cartesianism from more or less Cartesian positions. Thus, some varieties of Aristotelianism in the seventeenth century prepared the ground for the acceptance of Cartesianism and the eventual attempts at their reunification.

Part I closes with a key seventeenth-century question about the substance of the heavens: whether astronomical novelties such as sunspots and comets necessitate a significant change in cosmological theory. In Chapter 5, I examine Parisian cosmology during the seventeenth century in order to demonstrate the resiliency of traditional Aristotelian cosmology against the new astronomy of Galileo and Descartes. The authors surveyed include both Catholic and Protestant textbook writers at Paris and at Jesuit and non-Jesuit colleges around Paris (Jacques du Chevreul, Bouju, Pierre du Moulin, René de Ceriziers, Antoine Goudin, Jean Duhamel, Jacques Grandamy, and others).

Part II, "Debate and Reception," centers on the contrasts that Descartes drew between his philosophy and that of others, the controversies in which he was personally involved, and the controversy that Cartesianism caused in the second half of the seventeenth century: condemnations at Louvain, various French censures, and a formal dispute among minor French philosophers. Chapter 6 discusses the significance of some references in Descartes's correspondence to Sebastian Basso, an early seventeenth-century atomist. Initially, in October 1629, Descartes wrote to Marin Mersenne that he agreed with Basso about rarefaction but disagreed with him about the ether. Then, a year later, Descartes called Basso one of the *novatores* (along with Bernardino Telesio, Tommaso Campanella, Giordano Bruno, and Lucillio Vanini) in the context of an ill-tempered letter to Isaac Beeckman concerning what anyone can teach another. Basso, Descartes wrote, does not have anything to him, any more than anyone else, unless he can convince him by his reasons. Finally, in a letter to Constantijn Huygens, Descartes disavows Basso: he is only good for destroying Aristotle's opinion. Descartes denies that he shares Basso's intent and claims that he seeks only to establish something so simple and evident that everybody would agree with it. To make sense of these statements, I compare the doctrines of Descartes and Basso on the subject of rarefaction and the ether; this requires a discussion of their views on corpuscles and the void. But, above all, I contrast these doctrines with those of the Aristotle of the Scholastics at the start of the seventeenth century: Aristotelians such as Eustachius a Sancto Paulo and Scipion Dupleix, but especially Toletus and the Coimbrans, authors constantly cited by Basso and whom Descartes remembers reading in his youth.

Next, in Chapter 7, I analyze the exchanges between Descartes and the Jesuits of La Flèche on the mystery of the Eucharist. I regard these exchanges as Descartes's first steps toward a Cartesian Scholasticism. Contrary to the secondary literature on the subject, Descartes did not write about transubstantiation against his will—because he was forced to respond to Antoine Arnauld's questions about the Eucharist in the *Fourth Set of Objections* (1641)—or merely to flatter the Jesuits. Descartes himself freely raised these issues as early as 1630 and considered his explanation of the mystery of the Eucharist to be an excellent result of his philosophy (well before his exchange with Arnauld). For Descartes (as for others) two different aspects of the mystery required explanation: how, without using the scholastic doctrine of real accidents, the bread after transubstantiation might still look like bread to us (discussed in *Replies* IV); and how Christ may be really present in the consecrated bread (discussed in the *Letters to Mesland*). The chapter also demonstrates (what has not been previously noticed) that the Cartesian answer to the question of real presence is the standard seventeenth-century scholastic view.

The condemnations of Cartesianism by the authorities of Louvain are discussed in Chapter 8. In 1662, just a year before Descartes's works were placed on the *Index of Prohibited Books,* five propositions from Descartes's *Principles* were prohibited at Louvain. Specifically censured were Descartes's definition of substance in general, his rejection of substantial forms or real accidents, his doctrine that extension is the essential attribute of substance, his claim that the universe is indefinitely extended, and his rejection of multiple universes. Concentrating on the last two prohibited propositions, I investigate the scholastic background to these questions about the nature of the universe, that is, whether it is finite or infinite, single or plural.

I then consider two events in late seventeenth-century philosophy: the condemnation of Cartesianism by the church, the throne, and the university; and the noncondemnation of Gassendism by the same powers. What is striking about the two events is that both Cartesians and Gassendists accepted the same proposition deemed heretical: extension is the principal attribute of matter. Thus, what was sufficient to condemn Cartesianism was not sufficient to condemn Gassendism. To understand what is involved in condemnation, it seems clear, one has to pay close attention to the intellectual and/or social context and to rhetorical strategy, not just to the propositions condemned. In this case, some of the central propositions of corpuscularianism and the mechanical philosophy are involved.

Finally, I discuss Descartes's *cogito* in Chapter 10. But instead of asking logico-linguistic questions about it, I examine the critiques it received by seventeenth-century philosophers, in part for what they can tell us about these philosophers and in part for what they can reveal about Descartes and the *cogito* itself. First, it should be emphasized that critiques of the *cogito* were relatively rare, though not nonexistent, and that attending to them to the exclusion of the other interests of seventeenth-century philosophers might be misleading. After briefly reviewing the well-known critiques of the *cogito* by Hobbes and Gassendi, by the anonymous objectors of the *Sixth Objections,* and by the Jesuit Pierre Bourdin—together with Descartes's replies—I then examine the exchanges among Pierre-Daniel Huet, Pierre-Sylvain Régis, and Jean Duhamel. I argue that the seventeenth-century critiques of the *cogito* (especially those of Huet, Duhamel, and the inquisitors of the College of Angers) are very similar: all reject the *cogito* as a principle of knowledge or as science, properly speaking, because such a principle, according to the *Posterior Analytics,* must be a "commensurate universal," a proposition whose predicate belongs essentially to every instance of its subject. The *cogito,* thus, does not fit the scholastic model for pure scientific knowledge at all. It is neither universal nor necessary, but singular and contingent. Moreover, it is not a principle but an argument, and even a defective one; either it is dependent upon an unspecified major premise or it begs the question. And if it is an argument, it cannot be a principle of knowledge: an argument cannot itself be a principle. The rejection of the *cogito* in these critics is also linked with their prior rejection of Cartesian doubt.

These chapters all deal with Descartes's relations with the last Scholastics. The seventeenth-century collegiate course on physics (or the science of natural things) would have covered a variety of topics, from the order of the sciences to the materials of Aristotle's *Physics,* that is, principles (matter and form), causation (including exemplar causation), infinity, place, time, void, and motion; to the *De caelo, De generatione et corruptione,* and *Meterologica,* that is, the substance of the heavens, the elements (levity and gravity), meteors, comets, and other meteorological phenomena; to the *De anima,* that is, various souls, the senses (including internal and common senses), and other faculties of the soul—imagination, memory, appetition, understanding, will, memory. The course on Metaphysics (or supernatural science) would have begun with such topics as the object or subject of metaphysics, the question of whether the other sciences are subalternated to it, principles of metaphysics (being, existence and essence, cause and principle, archetypes or ideas), and transcendentals (unity, quantity, principle of individuation, truth and falsity, good and evil). It would have continued with mixed metaphysical-theological topics, such as whether man is created with knowledge, whether the soul is immortal and whether the separated soul retains its faculties. In my attempt to illuminate Cartesian philosophy by examining its context, analyzing some debates, and surveying some controversies, I hope to have touched upon most of the topics and themes shared by Cartesian and late scholastic philosophy.

[1]

Descartes among the Scholastics

For most readers of Descartes, the topic of Descartes's relations with the Scholastics brings to mind his disparaging comments about the philosophy he was taught: "*in my college days* I discovered that nothing can be imagined which is too strange or incredible to have been said by some philosopher."[1] Descartes, in the *Discourse on Method,* seemed to find little worthwhile in his education, including his education in scholastic philosophy and sciences; at best, "philosophy gives us the means of speaking plausibly about any subject and of winning the admiration of the less learned," and "jurisprudence, medicine, and other sciences bring honors and riches to those who cultivate them";[2] but "there is still no point in [philosophy] which is not disputed and hence doubtful" and, "as for the other sciences, insofar as they borrow their principles from philosophy . . . nothing solid could have been built upon such shaky foundations."[3]

Obviously, the Descartes of the *Discourse* represented himself as dissatisfied with school learning in general. When reading his correspondence, however, one can catch a glimpse of a different Descartes. In 1638, approximately a year after the publication of the *Discourse,* Descartes wrote a letter responding to a request for his opinion about adequate schooling for the correspondent's son. In the letter, Descartes attempted to dissuade the

---

1. AT 6:16; CSM 1:118; emphasis added. The statement is ambiguous, of course: it can be interpreted to mean that Descartes learned the Ciceronian phrase or that he had come to realize the matter himself. The pronouncements of the *Discourse* are formulae that echo standard skeptical assertions; for the literary background to the *Discourse,* see Gilson 1925. Still, the point is that disagreement about philosophical matters, and even the strangeness of philosophical positions, are part of the common knowledge shared by Descartes.
2. AT 6:6; CSM 1:113.
3. AT 6:8–9; CSM 1:115.

[7]

correspondent from sending his son to school in Holland. According to Descartes, "there is no place on earth where philosophy is better taught than at La Flèche,"[4] the Jesuit institution in which he was educated. Descartes gave four reasons for preferring La Flèche. First, he asserted, "philosophy is taught very poorly here [in Holland]; professors teach only one hour a day, for approximately half the year, without ever dictating any writings, nor completing their courses in a determinate time." Second, Descartes advised, "it would be too great a change for someone, when first leaving home, to study in another country, with a different language, mode of living, and religion"; La Flèche was not far from the correspondent's home, and "there are so many young people there from all parts of France, and they form such a varied mixture that, by conversing with them, one learns almost as much as if one traveled far." Descartes then praised as a beneficial innovation the "equality that the Jesuits maintain among themselves, treating in almost the same fashion the highest born [*les plus releuez*] and the least [*les moindres*]." Most important, Descartes asserted that although, in his opinion, "it is not as if everything taught in philosophy is as true as the Gospels, nevertheless, because philosophy is the key to the other sciences," he believes that "it is extremely useful to have studied the whole philosophy curriculum, in the manner it is taught in Jesuit institutions before undertaking to raise one's mind above pedantry, in order to make oneself wise in the right kind [of philosophy]."[5]

Of course, preferring La Flèche to a Dutch educational institution is not the same as giving an unqualified endorsement to La Flèche. On the other hand, some of Descartes's pronouncements, especially his last assertion, do seem inconsistent with those of the *Discourse*. How can the Descartes of the *Discourse* recommend learning scholastic philosophy as preparatory to the sciences and to his own philosophy? Is not the study of scholastic philosophy antithetical to the Cartesian project to cleanse oneself of the effects of years of dependence on the senses? Would not the study of scholastic philosophy merely reinforce those bad habits? Still, Descartes's advice in his letter seems open and frank, and Descartes's first three assertions in the letter correlate very well with what one can discover to have been the case in seventeenth-century Jesuit education.

Descartes was right in suggesting that students would have been taught more philosophy, and would have been taught it more rigorously at La Flèche than at a Dutch college or university. The philosophy curriculum at La Flèche is fairly well known, and the daily routine of its students well doc-

---

4. AT 2:378.
5. AT 2:378.

umented.[6] At La Flèche, as in other Jesuit colleges of the time,[7] the curriculum in philosophy would have lasted three years (the final three years of a student's education, from about the age of fifteen on). It would have consisted of lectures, twice a day in sessions lasting two hours each, from a set curriculum based primarily on Aristotle and Thomas Aquinas. During Descartes's time, the first year was devoted to logic and ethics, consisting of commentaries and questions based on Porphyry's *Isagoge* and Aristotle's *Categories, On Interpretation, Prior Analytics, Topics, Posterior Analytics,* and *Nicomachean Ethics*. The second year was devoted to physics and metaphysics, based primarily on Aristotle's *Physics, De caelo, On Generation and Corruption,* book 1, and *Metaphysics,* books 1, 2, and 11.[8] The third year of philosophy was a year of mathematics, consisting of arithmetics, geometry, music, and astronomy, including such topics as fractions, proportions, elementary figures, techniques for the measurement of distances and heights, trigonometry, gnomics, geography and hydrography, chronology, and optics.[9] The students would have been expected to study their professors' lectures thoroughly. Their daily routine would have included a number of hours of required study time. They would have had to show their work to a prefect daily and to repeat materials from their lectures to a *repetitor;* their learning would have been tested in weekly and monthly oral disputations in front of their professors and peers.

Descartes was not exaggerating when he asserted that the student population of La Flèche was diverse, geographically and otherwise. La Flèche accepted boys from all corners of France and from all walks of life. During Descartes's days, its boarders numbered approximately one hundred, and it taught, in addition, about twelve hundred external, or day, students. Moreover, the equality of treatment practiced by the Jesuits, and referred to by Descartes, does appear to be an innovation in the context of seventeenth-century France; it is verifiable by available documents. The sons of the most humble families lived in the same rooms as those of the most exalted. When arriving at La Flèche, one checked one's sword in the armory. "Without a sword, a gentleman forgot his birth; there would be no distinction between

---

6. For more information concerning La Flèche and its curriculum, consult Rochemonteix 1889; a popular exposition of the same material can be found in Sirven 1987.

7. For other colleges, as well as for general Jesuit educational theory, consult Wallace 1984, *Monumenta Paedagogica Societatis Jesu* 1901, and Dainville 1987; also Brockliss 1981, 1987.

8. Later, the second year became the year of physics and mathematics, with the third year being devoted to metaphysics.

9. See, for example, Gaultruche 1656, a good exemplar for what would have been taught in mathematics at La Flèche, given that Gaultruche was a Jesuit who taught mathematics at La Flèche and Caens.

nobility, bourgeois, etc."[10] There is even the case of Jean Tarin, one of Descartes's contemporaries, born in Anjou in 1586, who came to La Flèche "in poverty, with feet bare, and nothing but an undershirt and a bag of nuts and bread"; true, he was first a kitchen assistant and sweeper of classrooms for about four years, but then he became lackey to the young Comte de Barrant, who gave him the means and leisure to study. In 1616 he became professor of grammar at the Collège Harcourt, Paris, and in 1625 he became its rector.[11]

One can only conclude that the attitude toward scholastic education in philosophy displayed by Descartes in some of his correspondence more nearly represents his own views on the matter; at the very least, the letter to the anonymous correspondent about his son's education should provide one with a corrective for interpreting the negative views about scholastic education found in the *Discourse*.[12]

Descartes's Request for Objections: The Letters to Noël

Another letter written by Descartes about the time of the publication of the *Discourse* also casts doubt upon the reliability of any literal reading of that work. During June 1637 Descartes wrote to one of his old teachers, sending him a copy of the newly published *Discourse*. As Descartes put it, he sent the volume as a fruit that belongs to his teacher, *whose first seeds were sown in his mind by him,* just as he also owed to those of his teacher's order the little knowledge he had of letters.[13]

Now, it is true that Descartes sent copies of the *Discourse* to a great number of people: close friends, the nobility, various intellectuals, Jesuits, and

---

10. Rochemonteix 1889, 2:27.
11. Rochemonteix 1889, 2:25–27. Similarly, Marin Mersenne, Descartes's principal correspondent, was one of the students of humble origins who studied at La Flèche and played a role in the intellectual life of the seventeenth century. For Mersenne's intellectual biography, see Lenoble 1943 or Dear 1988.
12. As I have already indicated, it is difficult to reconcile Descartes's enthusiasm for La Flèche with his attitude on scholastic education in the *Discourse*. Of course, Descartes is merely stressing the academic rigor of the teaching, the discipline, and the social ethos of La Flèche; on the face of it this is quite compatible with the *Discourse* thesis that the subjects taught there were not much use. But why should one recommend a more rigorous school over a less rigorous one when what is taught more rigorously is of little use? This question becomes more pressing when one realizes that, as early as 1634, Regius (Chair of Medicine and, after 6 September 1638, extraordinary Professor at Utrecht) was already giving private lessons on Cartesian philosophy and physics, having been taught it by Reneri, Descartes's friend and earliest supporter in the Netherlands. It is one thing to recommend La Flèche as the best of a sorry lot, but another to recommend it over Utrecht, where one might be taught Cartesian philosophy.
13. AT 1:383.

others.¹⁴ It is also true that Descartes indicated in the letter that he had not kept in touch with his old teachers after he left La Flèche: "I am sure that you would not have retained the names of all the students you had twenty-three or twenty-four years ago, when you taught philosophy at La Flèche, and that I am one of those who have been erased from your memory."¹⁵ Moreover, the attempt to promote his works by making them the focus of discussion was already part of Descartes's strategy. When, in 1641, Descartes published his *Meditations on First Philosophy,* he did so with a series of *Objections* and *Replies* to the work. He had hoped to do the same thing with the earlier *Discourse.* In part 6 of the *Discourse,* Descartes announced: "I would be very happy if people examined my writings and, so that they might have more of an opportunity to do this, I ask all who have objections to make to take the trouble and send them to my publisher and, being advised about them by the publisher, I shall try to publish my reply at the same time as the objections; by this means, seeing both of them together the readers will more easily judge the truth of the matter."¹⁶ Thus, the letter Descartes wrote to his old teacher should be read from the above perspective; the letter was part and parcel of Descartes's strategy to promote discussions of his views. And, of course, Descartes did request objections from his teacher and from others of his order in the letter: "If, taking the trouble to read this book or have it read by those of your [order] who have the most leisure, and noticing errors in it, which no doubt are numerous, you would do me the favor of telling me of them, and thus of continuing to teach me, I would be extremely grateful."¹⁷ Still, it is curious to see the Descartes of the *Discourse* being so obsequious and sending his work to his teachers "as the fruit belonging to them, whose seed they sowed."

We do not have a response from Descartes's old teacher, but we can infer what he said, given that we have a second letter from Descartes to him, written in October 1637. Descartes thanked his correspondent for having remembered him and for giving his promise to have the book examined and objections forwarded. Descartes pressed his correspondent to append his own objections, saying that there are no objections whose authority would

---

14. See, for example, the letter of 14 June 1637 to Huygens (?), AT 1:387, in which Descartes indicates that, of the three copies of the *Discourse* enclosed, one is for the recipient of the letter, another for the Cardinal de Richelieu, and the third for the King himself.

15. AT 1:383. This sentence enables one to guess that the recipient of the letter is the Père Etienne Noël, Descartes's *repetitor* in philosophy, especially since Noël was rector of La Flèche in 1637. See Rodis-Lewis 1987, 190n; see also Rodis-Lewis, "Descartes aurait-il eu un professeur nominaliste?" and "Quelques questions disputées sur la jeunesse de Descartes," in Rodis-Lewis 1985, 165–81.

16. AT 6:75; CSM 1:149.

17. AT 1:383.

be greater, and none he desires more.[18] Descartes added that no one would seem to have more interest in examining his book than the Jesuits, since he did not see how anyone could continue to teach the subjects treated, such as meteorology, as do most of the Jesuit colleges, without either refuting what he had written or following it. At the end of the letter, however, Descartes seemed to recognize the reason why the Jesuits might not willingly take up his philosophy; he attempted to reply to the difficulty:

> Since I know that the principal reason which requires those of your order most carefully to reject all sorts of *novelties* in matters of philosophy is the fear they have that these reasons would also cause some changes in theology, I want particularly to indicate that there is nothing to worry from this quarter about these things, and that I am able to thank God for the fact that the opinions which have seemed to me most true in physics, when considering natural causes, have always been those which agree best of all with the mysteries of religion.[19]

It was clear to Descartes that the Jesuits' distaste of novelty, arising out of their desire to safeguard theology, would have been a stumbling block to friendly relations with them, for they would have rightly seen him as offering novelties. As in previous instances, Descartes seemed to understand his own situation fairly well; he seemed to have a clear grasp of Jesuit educational practices and objectives during the seventeenth century.

Jesuit Pedagogy in the Sixteenth and Seventeenth Centuries

During Descartes's lifetime, from his childhood at La Flèche to the 1640s, great changes in pedagogy were taking place. The Jesuits, following the example of the University of Paris, had reorganized their curricula.[20] They had undertaken extraordinary discussions and exchanged position papers, all of which ultimately led to the *ratio studiorum* of 1586, 1599, and 1616. As part of the self-consciousness about teaching, textbooks, both Jesuit and non-Jesuit, had undergone significant changes. Having decided to standardize their curricula, the Jesuits set out to write texts that reflected their curricular decisions. Early Jesuit textbooks presented Aristotle's texts in a most scholarly fashion; they were modeled after the great commentaries on

---

18. AT 1:454–56.
19. AT 1:455–56: CSMK, 75; emphasis supplied.
20. See Douarche 1970; also Brockliss 1987.

Aristotle's works, each volume treating a specific Aristotelian text (the *Physics, De anima, De caelo,* etc.), but presenting both the Greek text and Latin translations, together with Latin paraphrases (*explanationes*), leading to *quaestiones,* the treatment of standard problems relevant to particular texts, further subdivided into articles. Later Jesuit textbooks did the same, but deleted the Greek version of Aristotle's text. The textbooks of University of Paris professors deleted even Aristotle's Latin text: they simply strung together *quaestiones* in the order in which the text would have been presented, but did so for all Aristotelian sciences within the framework of the whole course of philosophy—Ethics and Logic, Physics and Metaphysics—in a single volume. The same held for popular contemporary presentations of the same materials in the French language.

But the contents of the textbooks was also a focus of discussion. There was a renaissance in Thomistic philosophy during the second half of the sixteenth century. For the duration of the Council of Trent (1545–63), Thomas's *Summa theologiae* was placed next to the Bible, on the same table, to help the council in its deliberations, so that it might derive appropriate answers. In 1567 Pope Pius V proclaimed Saint Thomas Aquinas Doctor of the Church. Saint Ignatius of Loyola, founder of the Jesuits, advised the Jesuits to follow the doctrines of Saint Thomas in theology.[21] Naturally, it would have been difficult to follow Saint Thomas in theology without also accepting much of his philosophy; and to follow Saint Thomas in philosophy would have required one to follow Aristotle as well. None of this was unexpected; Loyola's advice was made formal in the Jesuits' *ratio studiorum* of 1586: "In logic, natural philosophy, ethics, and metaphysics, Aristotle's doctrine is to be followed."[22] The flavor of the advice can be captured through a memorandum from the chief of the Order of Jesuits (Francisco Borgia) to the superiors of the order, written just after the end of the Council of Trent and imbued with the spirit of the Council and Saint Ignatius of Loyola's advice. I quote the memorandum in full:

That Which Must Be Held in Theology and in Philosophy

Let no one defend or teach anything opposed, detracting, or unfavorable to the faith, either in philosophy or in theology. Let no one defend anything against the axioms received by the philosophers, such as: there are only four

---

21. Rochemonteix 1889, 4:10, citing Loyola: "in theologia praelegendum esse S. Thomam." See also Hellyer 1996.
22. Rochemonteix 1889, 4:8n.

kinds of causes;[23] there are only four elements;[24] there are only three principles of natural things;[25] fire is hot and dry; air is humid and hot.[26]

Let no one defend anything against the most common opinion of the philosophers and theologians, for example, that natural agents act at a distance without a medium.[27]

Let no one defend any opinion contrary to common opinion without consulting the Superior or Prefect.

Let no one introduce any new opinion in philosophy or theology without consulting the Superior or Prefect.

Opinions That [Jesuits] Must Sustain, Teach, and Hold as True

Concerning God. God's power is infinite in intensity; He is a free agent according to the true philosophy. His Providence extends to all created beings in general, to each in particular, and to all human things; he knows all things present, past and future, according to the true philosophy.

---

23. The four kinds of causes, as given in Aristotle's *Physics* II, chaps. 3–10, are formal, material, efficient, and final; all four would be involved in a complete explanation of a change. For example, in the Aristotelian account of the reproduction of man, the material cause is the matter supplied by the mother, the formal cause is the specific form of man (that is, rational animal), the efficient cause is supplied by the father, and the final cause is the end toward which the process is directed.

24. Aristotle discusses the four elements in *De caelo* III and IV. The elements, that is, earth, water, air, and fire, are characterized by pairs of the contraries, hot and cold, moist and dry *(On Generation and Corruption* I); in Aristotle's theory of motion, the elements move naturally in a rectilinear motion, the first two elements having a natural downward motion, toward the center of the universe, whereas the second two have a natural upward motion, toward the periphery of the sublunar region. This creates a distinction between the sublunar world of the elements and the supralunar world of the heavens, whose ether moves naturally in a circular motion.

25. The three principles of natural things are form, matter, and privation, discussed by Aristotle in book 1 of the *Physics*. The form of a thing is its actuality, whereas the matter is its potentiality; privation is what the thing is not. For example, in a change from water being cold to being hot, heat is the form that the thing lacks, but it is water, the matter or subject, that gains the form and becomes hot (cold itself or the bare matter does not change). Change is the gaining or losing of forms; but some forms are essential and cannot be lost (for example, man cannot lose the form, rational animal, and remain man). Thus, a form is accidental when it confers a new quality to a substance already formed—heat, for example. On the other hand, a substantial form confers being; there is generation of a new being when a substantial form unites with matter, and real destruction when one separates from matter.

26. These "axioms" are sufficient to banish Stoic, Epicurean, and atomist philosophies. Epicureans and atomists account for change by the substitution or rearrangement of basic particles, or atoms, not by the replacement of forms in a matter capable of accepting various forms. Moreover, for an Epicurean or an atomist, the particles themselves would be more basic than the elements, and an insistence on four elements would go against Stoic cosmology.

27. This "common notion" is sufficient to reject the philosophy of non-Thomist Scholastics, such as Ockhamists. In his *Commentary on the Sentences* II, quaest. 18, Ockham accepts an account of magnetism as action at a distance, without the intervention of a medium, instead of accepting a medium as necessary for propagating a magnetic quality.

Concerning Angels. Angels are truly placed in categories and are not pure act, according to the true philosophy. They are in place and move locally from place to place, so that one should not hold that they are not in place and do not move, so also that their substance is present in some manner in one place and then in another.

Concerning Man. The intellective soul is truly the substantial form of the body, according to Aristotle and the true philosophy. The intellective soul is not numerically one in all men, but there is a distinct and proper soul in each man, according to Aristotle and the true philosophy.[28] The intellective soul is immortal, according to Aristotle and the true philosophy. There are not several souls in man, intellective, sensitive, and vegetative souls, and neither are there two kinds of souls in animals, sensitive and vegetative souls, according to Aristotle and the true philosophy.[29] The soul, whether in man or in animals, is not in fuzz or in hair. Sensitive and vegetative powers in man and animals do not have their subject in prime matter. Humors are, in some manner, part of man and animals. The whole being of composite substance is not solely in form, but in form and matter.

Varia. The predicables are five in number. Divine essence does not have a single subsistence common to three persons, but only three personal subsistences. Sin is a formal evil and a privation, not something positive. We are not causes of our own predestination.

Let all professors conform to these prescriptions; let them say nothing against the propositions here announced, either in public or in private; under no pretext, not even that of piety or truth, should they teach anything other than that these texts are established and defined. This is not just an admonition, but a teaching that we impose.[30]

Given the above, one might wonder whether Descartes's attempt to gain acceptance of his philosophy by the Jesuits was a quixotic endeavor. Descartes did try to indicate that his doctrines were not dangerous to the faith; but the Jesuits defined danger to the faith as any novelty in either theology or in philosophy, especially as it concerned the axioms and common opinions of Scholasticism. And Descartes would not have fared very well in this respect. He rejected the four causes, arguing that final causes are not

---

28. The target of this opinion is the Averroist doctrine of the numerical unity of intellective soul, that is, the doctrine denying the existence of individual souls and asserting that there is just one intellective soul.

29. The target of this opinion seems to be the Augustinian and Franciscan doctrines of the plurality of substantial forms. John Duns Scotus and William of Ockham held the thesis that man is a composite of forms (rational, sensitive, etc.), a thesis previously rejected by Thomas Aquinas, who argued that there is just one form or soul in man (the rational soul), which performs the functions that the other souls perform in lower beings.

30. Bibliothèque Nationale, mss. fonds Latins, no. 10989, fol. 87, as transcribed in Rochemonteix 1889, 4:4n–6n.

appropriate for natural philosophy.³¹ He set aside the four elements and held that there was only one kind of matter and that all its varieties could be explained as modifications of extensions.³² Moreover, Descartes did not accept the three Aristotelian principles of matter, form, and privation. Except for rational beings who have minds, Descartes discarded the doctrine of substantial forms.³³ Finally, though Descartes might have agreed that fire is hot and dry, and air is humid and hot, it would have been as phenomenological descriptions, and not as representing any basic reality; such statements would have been inconsistent with Descartes's mechanical philosophy, which required some kind of corpuscularianism, as well as the rejection of final causes and substantial forms (except for man's body as informed by a soul).

On the other hand, Descartes would have agreed with the common opinion that natural agents do not act at a distance without a medium.³⁴ Interestingly, Descartes could accept all the theological and philosophical opinions concerning God, angels, and man that Jesuits were required to sustain and defend, including the thesis that God's power is infinite in intensity,³⁵ that he is a free agent,³⁶ that the intellective soul in man is the substantial form of the body,³⁷ that the intellective soul is not numerically one in all men and that there is only one soul in man,³⁸ and that sin is a privation, not something positive.³⁹ The only notable exception was Descartes's denial of animal souls, both sensitive and vegetative.⁴⁰ Perhaps Descartes might have thought that his orthodoxy with respect to theological matters would have led to the acceptance of his philosophical novelties, once they were seen to harmonize with Catholic theological doctrines.

---

31. See Meditation IV, AT 7:55, and elsewhere.
32. Rule IV, AT 10:442, for example. If one wanted to draw Descartes closer to Aristotle—as does René le Bossu (1981 [1674], 286–87)—one could say that Descartes accepts three out of Aristotle's four elements, that is, fire, air, and earth. (See, for example, *Le Monde*, AT 11:25.) But that would be to disregard the important difference that Aristotle's elements are differentiated *qualitatively*, whereas there is only a *quantitative* difference among Descartes's elements. See also Chapter 4.
33. See *Principles* 4:art. 198, and elsewhere; Descartes does say (in a letter to Regius, AT 3:491–92) that he does not reject substantial forms overtly; he merely asserts they are not needed. The context of the assertion is an interesting letter in which Descartes counsels Regius to abstain from public disputes and from advancing novel opinions (that one ought to retain the old opinions in name, giving only new reasons).
34. Descartes is a mechanist and his world is a plenum. For the impossibility of void, see AT 4:329.
35. Meditation III, AT 7:45–50 (AT 9:32–40).
36. AT 1:152 and elsewhere.
37. For the doctrine that the numerical unity of a body does not depend upon its matter but its form, which is the soul, see the letter to Mesland, AT 4:346; CSMK, 278. See also Chapter 7.
38. AT 3:369–71; CSMK, 182.
39. AT 7:54; CSM 2:38.
40. AT 3:369–72; AT 6:56–59.

Perhaps also, during Descartes's time, there was a slightly more liberal interpretation given to Loyola's advice to follow Thomas. The traditional difficulty with the advice was that there were many divergent authorities, including those of the Church Fathers. This problem was handled straightforwardly in a memorandum by Claudio Aquaviva, fifth general of the Jesuits (1580–1615), to the superiors, written in order to express clearly the basic tenets underlying the *ratio studiorum* of 1586: "No doubt we do not judge that, in the teaching of scholastic theology we must prohibit the opinion of other authors when they are more probable and more commonly received than those of Saint Thomas. Yet because his authority, his doctrine, is so sure and most generally approved, the recommendations of our Constitutions require us to follow him *ordinarily*. That is why all his opinions whatever they may be . . . can be defended and should not be abandoned except after lengthy examination and for serious reasons." This interpretation of Loyola's advice drew a fine line between following Thomas's opinions *ordinarily* and abandoning them for extraordinary reasons, after lengthy examination. Surely, Descartes would have thought that he had abandoned Thomas's opinions only for serious reasons, after lengthy examination. Descartes's task would have been to demonstrate his reasons, to show that they are more probable. But Aquaviva's memorandum continued: "One should have as the primary goal in teaching to strengthen the faith and to develop piety. Therefore, no one shall teach anything not in conformity with the Church and received traditions, or that can diminish the vigor of the faith or the ardor of a solid piety." Aquaviva's intent was clear. The primary goal in teaching is the maintenance of the faith, and nothing should be allowed to interfere with it. All teaching must conform to the faith; and since the received traditions are known to conform to the faith, they should be taught and novelties are to be avoided. The memorandum continued:

> Let us try, even when there is nothing to fear for faith and piety, to avoid having anyone suspect us of wanting to create something new or teaching a new doctrine. Therefore no one shall defend any opinion that goes against the axioms received in philosophy or in theology, or against that which the majority of competent men would judge is the common sentiment of the theological schools.
> Let no one adopt new opinions in the questions already treated by other authors; similarly, let no one introduce new questions in the matters related in some way to religion or having some importance, without first consulting the Prefect of studies or the Superior.[41]

41. Bibliothèque Nationale, mss. fonds Latins, no. 10989, as transcribed in Rochemonteix 1889, 4:11n–12n.

The prohibition against holding or teaching new doctrines, against adopting new opinions, and even against introducing new questions in order not to diminish faith in any way would surely have made it difficult, if not impossible, for Descartes to have had his views accepted. Descartes's opinions went against many of the axioms received in philosophy. It would have been too optimistic an assessment to think that he might have gained acceptance with a majority of competent men in the theological schools.

Still, as conservative as the Jesuit practices seem, there was always the possibility that new doctrines might come to be accepted, especially those that did not seem to threaten the faith, those that appeared distant from theological matters. It is almost paradoxical that an order so outwardly conservative about philosophy and theology, with a pedagogy that rejects novelty, would have been able to produce novel works in meteorology, magnetic theory, geology, and mathematics.[42] On the other hand, the reasons why Jesuits avoided novelties were not dogmatic but prudential. One might therefore have expected rigid adherence to official positions, with respect to doctrines considered dangerous to piety, combined with some tolerance of doctrines considered nonthreatening.

Just such a strange mix of conservative and progressive doctrines can often be observed. Here, for example, are some doctrines from a public thesis in physics by a student at La Flèche, Jean Tournemine, in 1642.[43] In the section about the world and the heavens we are told that "the stars and firmament are not moved by an internal principle, but by intelligences." The thesis appears to be the rejection of some progressive elements of scholastic physics that could have blazed a path for the principle of inertia.[44] On the other hand, we are also told that "Apostolic authority teaches us that there are three heavens. The first is that of the planets, whose substance is fluid, as shown by astronomical observations; the second is the firmament, a solid body as its name indicates; and the third is the empyrean, in which the stars are specifically distinct from the heavens." This odd theory of the heavens breaks from the Aristotelian-Ptolemaic account of the heavens, fashionable in the seventeenth century, itself a modification of the Aristotelian system of homocentric spheres, adding Ptolemaic three-dimensional epicycles and eccentrics.[45] It is clearly at odds with Aristotelian

---

42. Cf. Heilbron 1979.
43. Joannes Tournemyne (La Flèche, 1642), as edited in Rochemonteix 1889, 4:365–68.
44. Including the rejection of fourteenth-century scholastic doctrines such as a circular impetus for the heavens. See Oresme 1968 and Albert of Saxony 1567.
45. As depicted, for example, in Eustachius a Sancto Paulo 1629, pt. 3, p. 96. It is interesting to note that "Apostolic authority" is invoked for the theory. Cf. Bellarmine's *Louvain Lectures* (1984).

principles about the heavens; the hypothesis of a fluid first heaven (and the theory as a whole) appears more suitable for the Tychonic scheme.[46]

Concerning the elements, it is asserted that "from the definition of element, it is obvious that four are to be posited, that is, earth, water, air, and fire, neither more nor less" and "heat, cold, wetness, and dryness are primary active qualities." These are extremely rigid assertions about the scholastic doctrine that seemed most under fire in the seventeenth century, especially the statement that the definition of "element" requires exactly four elements.[47] We are also told (as expected) that "the system of Copernicus on the daily rotation of the earth and its revolution around its own center, which is the immobile sun, is false and foolhardy"; but we are told that "none of the popular experiments are sufficient to assail it." This last admission seems to be very progressive (depending upon the reference to "popular experiments"), since it seems to indicate the acceptance of the relativity of motion.[48]

There is a palpable tension between the intellectual vigor of the new Order of the Jesuits setting up a whole new educational system and the attempt to reject novelty. This tension is evident even in an important event in which the young Descartes must have participated, the first memorial celebration of the death of Henry IV, the patron of La Flèche, on 4 June

---

46. The opposition between fluid and solid indicates that the thesis is not a version of the homocentric spheres made fluid. See Grant 1987. The reason why this theory of the heavens seems to be Tychonic is that solidity is attributed to the firmament, or the outermost heavenly body, containing the fluid universe of the planets. Fluidity is attributed to the world of the planets because of "astronomical observations." This seems to allude to the kind of observations of comets and novas that Tycho de Brahe used to argue against the solidity of planetary heavenly spheres. The Tychonic system, in which the earth was the center of the universe, with the planets revolving around the sun as their center, was a perfect compromise between the old Aristotelian-Ptolemaic system and the heliocentric Copernican system; it did not require a new physics for the motion of the earth. It did require, however, a fluid planetary heaven, since the paths of some planets intersected. Descartes discusses astronomical systems, including Tycho's in *Principles* III, articles 16–19, 38–41.

47. See Reif 1969.

48. It is difficult to tell what exactly was argued by the student in his thesis. But there were many "popular experiments" at the time claiming to refute Copernican astronomy; for example, cannon balls fired the same distance east and west were used as evidence against the rotation of the earth required by the Copernican system. According to modern principles of physics, these results cannot be counted against the rotation of the earth, so that the student's admission that "popular experiments" cannot defeat Copernicanism is interesting. During the same period, defenders of Copernicanism, such as Gassendi and Mersenne, used similar experiments in defense of Copernicanism: a stone falling from the mast of a moving ship falls parallel to the mast (Gassendi 1642), reported by Mersenne (in his 1644). It should also be pointed out that calling the Copernican system "false and foolhardy" is less harsh than calling it "foolish and absurd in philosophy and formally heretical," as did the Church in 1616. See below for Descartes's reaction to the Church's condemnation of Galileo's heliocentrism in 1633.

1611. For the occasion, the students of La Flèche composed and performed verses. The compositions were published for posterity as *Lacrymae Collegii Flexiensis* (La Flèche, 1611). One of the poems has the unlikely title "Concerning the Death of King Henry the Great and the Discovery of Some New Planets or Wandering Stars around Jupiter Noted the Previous Year by Galileo, Famous Mathematician of the Grand-Duc of Florence":

> France had already shed so many tears
> For the death of her King, that the empire of the waves,
> Heavy with water, ravaged the flowers of the earth,
> Threatening the whole world with a second flood;
>
> When the sun, making its rounds
> Around the universe, moved by the near disaster
> Which hastened toward us in its wandering path
> Spoke to her in this manner, amidst her pain:
>
> France, whose tears for the love of your Prince
> Do harm all other provinces by their excess,
> Stop grieving over your empty tomb;
>
> For God having taken him wholly from the earth
> He now illuminates the sky of Jupiter,
> That he may serve as a heavenly torch for all mortals.
>
> [La France avait déjà repandu tant de pleurs
> Pour la mort de son Roy, que l'empire de l'onde
> Gros de flots ravageait à la terre ses fleurs,
> D'un déluge second menaçant tout le monde;
>
> Lorsque l'astre du jour, qui faisait la ronde
> Autour de l'Univers, meu des proches malheurs
> Qui hastaient devers nous leur course vagabonde
> Lui parla de la sorte, au fort de ses douleurs;
>
> France de qui les pleurs, pour l'amour de ton Prince,
> Nuisent par leur excès à toute autre province,
> Cesse de t'affliger sur son vide tombeau;
>
> Car Dieu l'ayant tire tout entier de la terre
> Au ciel de Jupiter maintenant il esclaire
> Pour servir aux mortels de céleste flambeau.][49]

The sonnet has little literary merit, but in it the reader is treated to the image of the sun revolving around the earth, taking pity on the sorrow of

---

49. In Rochemonteix 1889, 1:147n–148n.

the French people for the loss of their king, and offering them a new torch—the new stars around Jupiter. It combines a naive, poetic view of the sun with an announcement of Galileo's discovery of the moons of Jupiter during the previous year.[50] The poem suggests that the students at La Flèche were made aware of the discovery, but perhaps not its significance, its use as an argument for the Copernican system and against the Aristotelian.

Early Objections and Replies: The Morin Correspondence

Descartes's request for objections and his sending out of copies did not bear much fruit. Early on, Descartes was uncertain whether he would receive a favorable reaction from the Jesuits. He wrote to Huygens: "As for my book, I do not know what opinion the worldly people will have of it; as for the people of the schools, I understand that they are keeping quiet, and that, displeased with not finding anything in it to grasp in order to exercise their arguments, they are content in saying that, if what is contained in it were true, all their philosophy would have to be false."[51] But he was hopeful; in the same letter he wrote:

> I have just received a letter from one of the Jesuits at La Flèche, in which I find as much approbation as I would desire from anyone. Thus far he does not find difficulty with anything I wanted to explain, but only with what I did not want to write; as a result, he takes the occasion to request my physics and my metaphysics with great insistence. And since I understand the communication and union that exists among those of that order, the testimony of one of them alone is enough to allow me to hope that I will have them all on my side.[52]

Ultimately, Descartes received a number of responses; among them was one from Libertius Fromondus, an anti-atomist, one from Plempius, a student of Fromondus, and a third from Jean Baptiste Morin, a progressive Aristotelian.[53] Fromondus treated Descartes as an atomist and sent him a tract against Epicureans and atomists that he had written earlier; but he did

---

50. Galileo, *Siderius Nuncius*.
51. AT 2:48.
52. AT 2:50.
53. Descartes was asked by Mersenne whether foreigners formulated better objections than the French. Descartes replied that he did not count any of those received as French other than Morin's objections. He referred to a dispute with Petit, which he dismissed, saying that he did not take Petit seriously but simply mocked him in return. (For more on the exchange between Descartes and Petit, see Jean-Luc Marion, "The Place of the *Objections* in the Development of Cartesian Metaphysics," in Ariew and Grene 1995, 7–20.) Descartes then listed the objections of the foreigners: Fromondus from Louvain, Plempius, an anonymous Jesuit from Louvain, and someone from the Hague. AT 2:191–92: CSMK, 105.

not respond to Descartes's reply. Descartes wrote to Huygens concerning the exchange: "As for Fromondus, the small disagreement we had is not worth your knowing about.... In any case, this dispute between us was more like a game of chess; we remained good friends."[54] The correspondence with Plempius was lengthier, with many letters debating biological matters, such as the theory of the circulation of the blood, going back and forth.[55] But the most interesting exchange was that between Descartes and Jean Baptiste Morin, who wrote to Descartes on 22 February 1638, with some comments on astronomy and Descartes's theory of light.

In the exchange, Morin engaged Descartes in some provocative metaphilosophical issues. First, Morin complained that Descartes, whose mind was used to the most subtle and lofty speculations of mathematics, closed himself off and barricaded himself in his own terms and manners of speaking, in such a way that he seemed at first almost impregnable.[56] He then stated,

> However, I do not know what to expect from you, for some have led me to believe that, if I used the terms of the schools, even a little, you would instantly judge me more worthy of disdain than of reply. But, reading your discourse, I do not judge you the enemy of the schools, as you are depicted.... The schools seem only to have failed in that they were more occupied by speculation in the search for terms needed to treat things, than in the inquiry into the very truth of things by good experiments; thus they are poor in the latter and rich in the former. That is why I am like you in this respect; I seek the truth of things only in nature and do not place my trust in the schools, which I use only for their terms.[57]

Descartes's answer is interesting. First, he assured Morin that he did not try to close off and barricade himself in obscure terms as a defensive move, and that if he did make use of mathematical demonstrations, it is because they taught him to discover the truth, instead of disguising it.[58] He then stated, "As for my disdain for the schools that you've been told about, it can only have been imagined by people who know neither my habits nor my dispositions. And though, in my essays, I made little use of terms known only by the learned, that is not to say that I disapprove of them, but only that I wanted to make myself understood also by others."[59] Later on, in the

---

54. AT 2:49. The correspondence between Descartes and Fromondus as well as that between Descartes and Morin is discussed in Garber 1988.

55. The correspondence between Descartes and Plempius is discussed in Grene 1995; it was not always a pleasant exchange.

56. AT 1:40.

57. AT 1:541.

58. AT 2:200–201; CSMK, 108.

59. AT 2:201–2.

same letter, defending himself against one of Morin's objections, Descartes accepted some scholastic distinctions. Trying to impress Morin with his knowledge of scholastic terminology, he peppered his letter with such terms: "I freely use here the terms of the schools in order that you do not judge that I disdain them."[60] He insisted on responding to Morin *in forma;* he threw in some terms and phrases from scholastic disputations, such as *distinguo, concedo totum, nego consequentiam,* and he even suggested that he was taking the term "infinite" *syncategorematice* "so that the schools would have nothing to object to in this matter."[61]

There is an amusing reply to Descartes's letter in a marginal comment to a letter from Mersenne to Descartes:

> You so reassured and enriched us by the excellent replies you made to Mr. Morin and me, that I assure you, instead of the 38 sols of postage on the package, seeing what it contained, I would have willingly given 38 ecus. We read the reply together; and Mr. Morin found your style so beautiful that I advise you never to change it. For your analogies and your curiosities satisfy more than what all others produce.... Moreover, you succeeded very well, in the reply to Mr. Morin, by showing that you do not disdain, or at least, you are not ignorant of Aristotle's philosophy. That is what contributed toward the increase in esteem Mr. Morin testifies as having for you. It is also what I assure those who, deceived by the clarity and precision of your style—which you can lower to make yourself understood by the common man—believe that you do not understand scholastic philosophy at all; I let them know that you understand scholastic philosophy just as well as the masters who teach it and who seem most proud of their own ability.[62]

The greater esteem Morin felt for Descartes did not prevent him from sending a second letter, in the style of Descartes's response, still objecting about the uses of terms, etc. Descartes responded to the letter, but with less enthusiasm. Morin wrote a third letter, but Descartes stopped the correspondence there. Descartes wrote to Mersenne, "I will not reply to Mr. Morin, since he does not want me to. Also, there is nothing in his last letter

---

60. AT 2:205.
61. AT 2:205–7. *In forma* means in logical form; *distinguo, concedo totum,* and *nego consequentiam* mean "I distinguish," "I concede totally," and "I deny the consequence." "Taking the term 'infinite' syncategorematically" alludes to medieval refinements of Aristotle's doctrine on potential infinity (versus actual infinity) from *Physics* III, chaps. 4–8. For more on infinity as a syncategorematic term, see Chapter 8.
62. AT 2:287. It is difficult to believe that Mersenne is being straightforward in his marginal comment—that he believes Descartes to understand scholastic philosophy as well as the masters who teach it. Mersenne himself can be said to understand scholastic philosophy very well, as his writings demonstrate, and to have kept up with the various disputes. On the other hand, as we shall see, even Descartes is aware of his own shortcomings in this respect, aware that he has not read scholastic philosophy for the last fifteen years or so.

that gives me the occasion to reply with something useful; between us, it seems to me that his thoughts are now farther from mine than they were at the beginning, so that we will never come to any agreement."[63]

The anticipated objections and replies to the *Discourse* seem to have failed completely. When Fromondus bothered to respond, it was not to start a dialogue. Worse yet, when a dialogue was started, as in the case of Morin, it did not result in any meeting of minds. How could Descartes have expected to succeed in winning over the more conservative members of the intellectual community, including those with a specific intellectual agenda, such as the Jesuits, when he could not convince someone like Morin of his views? Morin, a renowned optical theorist, astrologer to the king, and professor of mathematics at the Collège de France, at least styled himself a progressive thinker. "I am like you," he said to Descartes, "in that I seek the truth of things in nature and do not place my trust in the schools, which I use only for their terms." It is true that Morin was anti-atomist and anti-heliocentrist, as were the conservatives, but he was a mathematics professor at the Collège de France, not a theologian or faculty member of a Jesuit college, and at least he was willing to entertain a debate. The exchanges with Fromondus and Morin could not have pleased a philosopher who held that when someone has the truth he cannot fail to convince his opponents.[64]

The Bourdin Affair and the Eustachius Project

Descartes's relations with the Jesuits took a new turn in 1640. On 30 June and 1 July, a professor at Clermont, the Jesuit college in Paris, held a public disputation in which his student, a young noble named Charles Potier (who later became a Cartesian), defended some theses; among the theses were three articles concerning Descartes's theory of subtle matter,[65] reflection, and refraction. The professor, Father Bourdin, composed a preface to the thesis, called a *velitatio* (skirmish), which he delivered himself. Mersenne attended the disputation and defended Descartes. He apparently chastised Bourdin for having attacked Descartes publicly without having sent Descartes his objections; Mersenne then forwarded Descartes the

---

63. 15 November 1638, AT 2:437.
64. *Regulae* Rule II, AT 10:363; CSM 1:11.
65. Descartes's world is a plenum of subtle matter (ether, or first matter) whose action is used by Descartes to explain such diverse phenomena as gravitation and light. In the theses, Bourdin is complaining about Descartes's use of subtle matter for the propagation of light in *Optics* I, 5–7, "as a blind man can sense the bodies around him using his cane" (AT 6: 84: CSM 1:153).

velitatio, together with the three articles concerning Descartes's doctrines, as if they came from Bourdin himself.[66]

Descartes wrote to Mersenne on 22 July 1640, thanking him for the affection Mersenne showed for him "in the dispute against the theses of the Jesuits." He told Mersenne that he had written to the rector of the Collège de Clermont requesting that they address to him their objections against what he had written, "for he does not want to have any dealings with any one of them in particular, except insofar as it would be attested to by the order as a whole."[67] And he complained that the *velitatio* Mersenne sent him was "written with the intent to obscure rather than to illuminate the truth."[68] At the same time, Descartes wrote to Huygens, telling him, "I believe that I will go to war with the Jesuits; for their mathematician of Paris has publicly refuted my *Dioptrics* in his theses—about which I have written to his Superior, in order to engage the whole order in this dispute."[69]

The Bourdin affair degenerated, Descartes consistently referring to Bourdin's objections as *cavillations*.[70] The period of this dispute was a particularly difficult one for Descartes, since it was the time of his publication of the *Meditations*, his work on "First Philosophy," or metaphysics, which he had only sketched in the *Discourse* and which was certain to lead him into greater controversies, given that its content was closer to theology than was that of the *Discourse* with its appended *Essays* on physical and mathematical topics. The summer of 1640 was also the time when Mersenne was sending out Descartes's *Meditations* to intellectuals, requesting objections that would be published with the *Meditations*. Descartes even expected a set of objections from Bourdin himself.[71] One has to remember that this enterprise

66. Baillet 1691, 2:73. Bourdin was professor of humanities at La Flèche (1618–23), of rhetoric (1633), and mathematics (1634). He was sent to Paris, to the Collège de Clermont (later known as the Collège Louis-le-Grand) in 1635. By 1640, when Bourdin debated with Descartes, he had already published three books, *Prima geometriae elementa*, following Euclid (Bourdin 1639), *Geometria, nova methodo* (Bourdin 1640), and *Le cours de mathématique* (Bourdin 1661, first published circa 1631); he would shortly be publishing his fourth, *L'introduction à la mathématique* (Bourdin 1643). Bourdin's mathematics, like most Jesuit mathematics, can be characterized by its practical bent. This is made clear by Bourdin's *Cours de mathématique*, which contains materials on fortifications, terrain, military architecture, and sections on cosmography and the use of a terrestrial globe; it is also supported by the subject of his two posthumous publications: *L'architecture militaire ou l'art de fortifier les places regulières et irregulières* and *Le dessein ou la perspective militaire* (Bourdin 1655a and 1655b). See Ariew 1995.

67. AT 3:94.

68. AT 3:94. In another letter, Descartes tells Mersenne that he is shocked by the *velitatio* of Bourdin, for he does not have a single objection to anything Descartes has written, but rather attacks doctrines Descartes does not hold. AT 3:127–28.

69. AT 3:103; CSMK, 151.

70. That is, "quibbles" or "cavils." See AT 3:163, 184, 250, for example.

71. Bourdin wrote the *Seventh Objections*, which were not received by Descartes in time for the first printing of the *Meditations and Objections and Replies* but were included in the second printing.

would be crucial for Descartes if he expected to win his war against the Jesuits. The whole affair should be put into the context of the failure of the requested objections and replies to the *Discourse,* the unsuccessful correspondence with Morin, and the subsequent open hostilities with the Jesuits.

On 30 September 1640 Descartes wrote to Mersenne: "the cavils of Father Bourdin have resolved me to arm myself from now on, as much as I can, with the authority of others, since the truth is so little appreciated alone." In this context he told Mersenne that he will not travel that winter, since he is "expecting the objections of the Jesuits in 4 or 5 months," and he believes that he "must put himself in the proper posture to await them."[72] He then made an unusual request and an interesting revelation:

> As a result, I feel like reading some of their philosophy—something I have not done in twenty years—in order to see whether it now seems to me better than I once thought. Toward that end, I beg of you to send me the names of authors who have written textbooks in philosophy and who have the most following among the Jesuits, and whether there are new ones from twenty years ago; I remember only the Coimbrans, Toletus, and Rubius. I would also like to know whether there is someone who has written a summary of all of scholastic philosophy and who has a following, for this would spare me the time to read all their heavy tomes. It seems to me that there was a Chartreux or a Feuillant who had accomplished this, but I do not remember his name.[73]

The Scholastics Descartes remembered, the Coimbrans, Toletus, and Rubius, were all Jesuit textbook authors Descartes probably read at La Flèche. The Coimbrans (the Conimbricenses) were professors at the Colegio das Artes in Coimbra, Portugal, who published a series of encyclopedic commentaries on Aristotle's works between 1592 and 1598. The most noted of the Coimbrans was Petrus de Fonseca, who contributed to the *ratio studiorum* and who published separately his own commentaries on the *Metaphysics* and the *De anima.*[74] Franciscus Toletus was a professor at the Collegio Romano (1562–69) who published numerous commentaries on Aristotle's works, including an important *Logic* (1572), *Physics* (1575), and *De anima* (1575).[75] And Antonius Rubius taught philosophy in Mexico; he published commentaries on Aristotle's *Logic,* the *Logica mexicana* (1603), *Physics* (1605), *De caelo* (1615), and *De anima* (1611).[76]

We do not have Mersenne's reply, but presumably he identified Eustachius a Sancto Paulo as the Feuillant that Descartes remembered having

---

72. AT 3:184–85.
73. AT 3:185; CSMK, 154.
74. See C. H. Lohr 1975 and 1976. See also Schmitt and Skinner 1987, 814, 818.
75. See Lohr 1982 and Schmitt and Skinner 1987, 838.
76. See Lohr 1980.

written a summary of all of scholastic philosophy in one volume, since Descartes wrote in his next letter to Mersenne: "I have purchased the *Philosophy* of Brother Eustachius a Sancto Paulo, which seems to me to be the best book ever written on this matter; I would like to know whether the author still lives."[77] Eustachius a Sancto Paulo (Asseline) entered the Feuillants, a Cistercian order, in 1605 and was professor of theology at the Sorbonne. He published the *Summa philosophica quadripartita de rebus dialecticis, moralibus, physicis, et metaphysicis* in 1609. It was published repeatedly until 1648.[78]

We should make no mistake about the sense of Descartes's praise of Eustachius's *Summa* as "the best book ever written on this matter." In the same letter, Descartes says about the philosophy of the schools, "As for scholastic philosophy, I do not hold it as difficult to refute on account of the diversity of their opinions; for one can easily upset all the foundations about which they are in agreement among themselves; and that accomplished, all their particular disputes would appear inept."[79] This judgment was reinforced as Descartes read more scholastic textbooks, seeking a textbook as good as that of Eustachius but written by a Jesuit. Descartes told Mersenne, "I will also look at the Philosophy text of Mr. Draconis [that is, de Raconis], which I believe will be found here; for if he is more brief than the other and as well received, I will prefer it."[80]

Charles d'Abra de Raconis was born a Calvinist and converted to Catholicism. He taught philosophy at the Collège des Grassins and the Collège du Plessis, Paris. He then held a chair of theology at the Collège de Navarre, also in Paris. He published his *Summa totius philosophiae* in 1617, republishing it in parts and expanding it numerous times until 1651.[81]

Descartes wrote later: "I have seen the *Philosophy* of Mr. Raconis, but it is not as suitable for my design as that of Father Eustachius. And as for the Coimbrans, their writings are too lengthy; I would have wished wholeheartedly that they had written as briefly as the other, since I would have preferred to have dealings with the society as a whole, instead of a particular person."[82]

Descartes seems to have gained confidence as he read scholastic philosophy. He told Mersenne, "I thank you for the letter you have transcribed for

---

77. AT 3:232.
78. See Lohr 1976.
79. AT 3:231–32; CSMK, 156.
80. AT 3:234.
81. See Lohr 1974.
82. AT 3:251. Descartes does not mention one of the more interesting works in the genre, Dupleix 1627. Scipion Dupleix was more a historian than a philosopher, summarizing the school learning of his day as succinctly as possible, for an audience that is not comfortable with Latin, that is, an unschooled audience. Cf. Faye 1986.

me; but I find nothing useful in it, nor anything that seems as improbable to me as the philosophy of the schools."[83] He also informed Mersenne of his new project, the "design" to which he referred in the previously cited letter:

> My intent is to write in order a textbook of my philosophy in the form of theses, in which, without any superfluity of discourse, I will place only my conclusions, together with the true reasons from which I draw them—what I think I can do in a few words. And in the same book, I will publish an ordinary philosophy text [that is, a school text], such as perhaps that of Brother Eustachius, with my notes at the end of each question, to which I will add the various opinions of others and what one should believe about all of them, and perhaps, at the end, I will draw some comparisons between these two philosophies.[84]

Later he informed Mersenne that he had begun the project.[85] He wrote to others about it; he floated a trial balloon with the chief of the Jesuits, almost using the project as a threat, but also trying to determine the Jesuits' reaction to it. He even attributed the project to one of his unnamed friends.[86] But the project was soon aborted: "I am unhappy to hear about the death of Father Eustachius; for, although this gives me greater freedom to write my notes on his philosophy, I would nevertheless have preferred to do this with his permission, while he was still alive."[87] Descartes continued to use the project as a threat or bargaining chip with the Jesuits, but he no longer seemed willing to produce the work. He wrote to Mersenne, concerning a letter from Bourdin, "I believe that his Provincial sent it in order to ask you whether it is true that I am writing against them. . . . It is certain that I would have chosen the compendium of Father Eustachius as the best,

---

83. AT 3:256.
84. AT 3:233; CSMK, 157.
85. AT 3:259; CSMK, 161. The only new element added by Descartes concerned his hopes for the project, "that those who have not yet learned the philosophy of the schools will learn more easily from this book than from their teachers, because by the same means they will learn to scorn it, and even the least teachers will be capable of teaching my philosophy by means of this book alone" (AT 3:259–60). About a month later, Descartes was deep into the project, having just completed the first part of the *Principles* corresponding with the *Meditations*, and being able to compare the two: "I would be pleased to have as few distractions as possible, at least this year, since I have resolved to write my philosophy in such an order that it could easily be taught. And the first part, which I am now writing, contains almost the same things as the *Meditations* you have, except that it is written in a different style, and that what is written about at length in the one is abbreviated in the other, and vice versa" (AT 3:276).
86. AT 3:270.
87. AT 3:280. Descartes had previously indicated that he only wanted to do the project "with the writings of a living person and with his permission, which it seems to me I would easily obtain when my intention, to consider the one I chose as the best of all who have written on philosophy, will be known" (AT 3:234). But Descartes also added: "I have completely lost the intent to refute this philosophy; for I see that it is so absolutely destroyed by the establishment of mine alone that there is no need of another refutation" (AT 3:470).

if I wanted to refute someone; but it is also true that I have completely lost the intent to refute this philosophy; for I see that it is so absolutely and so clearly destroyed by means of the establishment of my philosophy alone, that no other refutation is needed."[88]

The Eustachius project is instructive for many reasons. One of the inferences one should draw from it is that Descartes was not well acquainted with scholastic philosophy in the period of his greatest work, during 1637–40. When he finally formulated his mature works, he departed either dramatically or by degrees from a scholastic tradition he no longer knew very well. Of course, Descartes was taught scholastic philosophy in his youth at La Flèche, but he abandoned his study of it for about twenty years, roughly between 1620 and 1640, and he picked it up again only in 1640 to arm himself against the expected attacks of the Jesuits. We should expect that Descartes was generally well versed in scholastic philosophy[89] only when writing his earliest works, the *Rules for the Direction of the Mind*, for example. (The remnants of Scholasticism in Descartes's mature works, the *Discourse* and the *Meditations*, might therefore be deceptive for the interpreter.)[90] Finally, from 1640 on, in the *Replies* to the *Objections* to the *Meditations* and in the *Principles of Philosophy*, Descartes relearned scholastic philosophy (and scholastic terminology) and began the process of reinterpreting his thoughts (or translating his doctrines) to make them more compatible with Scholasticism.[91] One can detect Descartes's subtle shifts in doctrine or terminology by contrasting his early and later writings—roughly, those before and after 1640.

---

88. AT 9:170. For Descartes's keeping open the option to write such a philosophy as a threat against the Jesuits, see AT 3:470, 480–81. One has to remember that there were political and social considerations in the project of comparing the two philosophies, the consequences of which were not lost on Descartes. For example, Descartes considered publishing his project in Latin and calling it *Summa Philosophiae* for tactical reasons in what he called "a scholastic war." He said to Huygens, "Perhaps these scholastic wars will cause my *Le Monde* to be brought into the world. I believe it would be out already, were it not that I would want first to teach it to speak Latin. I would call it *Summa Philosophiae*, so that it would be more easily introduced into the conversation of the people of the schools, ministers as well as Jesuits, who are now persecuting it and trying to smother it before its birth" (AT 3:523).

89. But probably only the scholastic philosophy represented by the Coimbrans, Toletus, and Rubius, that is, a sixteenth- and seventeenth-century neo-Thomism.

90. I think of the talk of a single, all-embracing Cartesian method—exemplified by Descartes's tree of philosophy, in which knowledge is an organic whole and all fields have the same method—opposed to that of Aristotle, for whom the different fields of human knowledge all have their own subject matter and appropriate methods, as the kind of interpretive doctrine that does not pay enough attention to Descartes's changing terminology from the 1620s to the 1640s. Descartes's tree is straight out of de Raconis's *Physica* and the doctrines about analysis and synthesis in *Replies* II seem to be consistent with seventeenth-century scholastic views, though the method in the *Rules* does seem directed against Aristotle and Aristotelians; see Ariew 1996.

91. For differences between Jesuit Scholasticism and non-Jesuit Scholasticism, see Chapters 2 and 4.

It is well known that Descartes refused to publish his *Le Monde* after being told of the condemnation of Galileo by the Catholic Church in 1633. The Church had declared the immobility of the sun to be foolish and absurd in philosophy and formally heretical, and the motion of the earth to merit the same censure in philosophy and to be at least erroneous in faith. Clearly, the Church was attempting to defend the faith, but it was also upholding a particular philosophy; the immobility of the earth and revolution of the sun around the earth were tenets of Aristotelianism. Descartes responded in characteristic style: "this has so astonished me that I almost resolved to burn all my papers, or at least not to let anyone see them. For I cannot imagine that Galileo, who is Italian and even well-loved by the Pope, as I understand, could have been made a criminal for anything other than having wanted to establish the motion of the earth."[92] In his *Le Monde* Descartes was clearly committed to the motion of the earth: "I confess that, if the motion of the earth is false, all the foundations of my philosophy are also. For it is clearly demonstrated by them. It is so linked to all parts of my treatise that I cannot detach it without rendering the rest defective."[93] So Descartes withheld publication and measured his public utterances on this issue. He avoided all discussion of it in the synopsis he gave of *Le Monde* in his 1637 *Discourse* and, when he finally took a public stance on the issue, in his 1644 *Principles* of *Philosophy*, it was to claim that "strictly speaking, the earth does not move."[94]

Descartes's philosophical progress on the motion of the earth seems to have resulted in a politically more tenable position. Descartes was pessimistic in *Le Monde* about the possibility of a definition of motion; he even ridiculed the Scholastics' definition: "To render it in some way intelligible, they have not been able to explain it more clearly than in these terms: *motus est actus entis in potentia, prout in potentia est*. For me these words are so obscure that I am compelled to leave them in Latin because I cannot interpret them."[95] For Descartes, the nature of motion is simpler and more intelligible than the nature of other things; it is used to explain other things—lines as the motion of a point and surfaces as the motion of a line, for example, instead of being explained by them. But, in the *Principles*, Descartes gave his own definition of motion, both in the ordinary sense of the word and in the strict sense, and even contrasted it with that of the Scholastics.[96]

92. AT 1:270–71; CSMK, 41.
93. AT 1:270–71; CSMK, 41.
94. *Principles* III, art. 28.
95. AT 11:39; CSM 1:94.
96. *Principles* II, arts. 24–25.

Similarly, Descartes criticized the related scholastic doctrine of place in his early works; he rejected the Scholastics' concept of intrinsic place[97] and parodied their concept of imaginary space.[98] But in the *Principles*, Descartes developed a doctrine of internal and external place clearly indebted to those he had previously rejected.[99] With respect to the scholastic concept of place, he asserted in the *Rules* (circa 1628): "When they define place as 'the surface of the surrounding body,' they are not really conceiving anything false, but are merely misusing the word 'place.'"[100] But by 1644, Descartes could go along with the misuse, defining space as internal place and relating it to external place. In *Principles* II, articles 10–12, he asserted that space, or internal place, does not differ from the corporeal substance contained in it, except in the way that we conceive of it; the same extension that constitutes the nature of body also constitutes the nature of space. On the other hand, in articles 13–15 he defined external place and its relation to space. For Descartes, "place or space do not signify a thing different from the body which is said to be in place, but only designate its size, shape and situation among bodies." To determine its situation among bodies, however, we must take into account other bodies we consider motionless. So we can define an external place, namely, the surface of the surrounding body and ultimately some supposedly motionless points in the heavens, as the fixed and determinate place for the motion of a body. The body then might simultaneously change and not change its place: it might change its external place (its situation) and not change its internal place (its extension or shape). Given that Descartes thought it impossible to discover any truly motionless points in the universe, he also thought that "nothing has any enduring, fixed and determinate place, except insofar as its place is determined in our minds." Thus, for Descartes, place, properly speaking, is internal place, or space, which is to be identified with the nature of body, that is, its extension, but we can mentally construct a situation, or external place, as the immobile reference for the motion of bodies.[101]

---

97. *Regulae*, AT 10:433–34; CSM 1:53.
98. AT 11:31.
99. Cf. *Principles* II, arts. 10–15. One can find these distinctions in part 3 of Eustachius a Sancto Paulo's *Summa*. See Chapter 2.
100. *Regulae*, Regula 13.
101. Descartes's choice of a relatively conservative, Aristotelian-inspired theory of space must have been a conscious decision. At the time, it was possible to choose from a number of non-Aristotelian concepts of space, originating from attempts to reestablish ancient views, such as Platonism and Epicureanism. Among the new Platonists were Giordano Bruno, Bernardino Telesio, and Tommaso Campanella. All three conceived of space as a container,

One can multiply such instances, but perhaps one more example might suffice to show that these instances are not limited to the more scientific aspects of Descartes's philosophy. One of the Cartesian philosophical doctrines under attack was the doctrine of material falsity. In the *Meditations* Descartes characterized material falsity as "occurring in ideas, when they represent non-things as things."[102] Descartes's example of material falsity was his idea of cold, which, though it is merely the absence of heat, represents cold as something real and positive. As Arnauld rightly pointed out in his *Objections* to the *Meditations*, "if cold is merely an absence, then there cannot be an idea of cold which represents it to me as a positive thing."[103] Descartes's response seems to have been a shift away from his initial position; that is, Descartes asserted in the *Replies* that the reason he called the idea of cold materially false was that he was unable to judge whether or not what it represented to him was something positive existing outside his sensation.[104] But there was also an interesting addition in his reply. Descartes seems to have used the occasion to show off his knowledge of scholastic philosophy in an ostentatious manner; the reply looks suspiciously similar to those given to Morin. Descartes, who did not usually cite sources, went out of his way to state that he did not worry about his use of material falsity, because Suarez defined material falsity in the same way in his *Metaphysical Disputations* (disp. 9, sec. 2, n. 4).[105] The response is even more curious, given that Descartes did not refer to Suarez anywhere else, even though his correspondents did refer to him. And Suarez's scholastic doctrine is yet a third notion of material falsity. Suarez's doctrine was basically an expansion of the Thomist doctrine that truth and falsity consist in composition and divisions.[106] Thus, material falsity as used by Suarez was about propositions, not ideas.

---

independent of bodies, but always occupied by bodies. Moreover, Francesco Patrizi maintained a more radical line in which three-dimensional space was to be thought of not as substance or accident, but as something subsisting in itself, inhering in nothing else. Space is the infinite, immobile container in which God placed bodies, filling some places, but leaving others empty.

102. AT 7:44; CSM 2:30.
103. AT 7:207; CSM 2:145.
104. AT 7:234; CSM 2:164. Wilson 1978, 115–16, argues that Descartes's reply to Arnauld is inconsistent with his doctrine in Meditation III.
105. Replies IV, AT 7:235.
106. Aquinas, *On Interpretation* I, lect. 1, n. 3. Here is Suarez, disp. 9, sect. 2, n. 4: "Composition and division can be found either in the apprehension of a concept alone, abstracting from any judgment, or in a conception which involves a judgment at the same time. In the former case we have said that complex truth is properly found in the composition of our judgments; and the same must therefore be said of falsity, for contraries are of the same kind. Hence no one thinks he is deceived or goes astray until he judges how many false compositions he apprehends. But since the apprehension that happens without judgment regularly comes about by concepts of words, rather than things . . . (for in the composition of words there is falsity just as in the composition of signs), it can be admitted that in such apprehen-

There seems to have been some vacillation in Descartes's mind between the material falsity of an idea as representing being as nonbeing and as having so little content that we cannot tell whether it represents something or not; but Descartes aggravated the apparent vacillation with an uncharacteristic and surprising reference to Suarez on material falsity as arising from composition and division. In the end, the doctrine of material falsity seems to have disappeared entirely. It did not recur in the *Principles*, possibly having been replaced by Suarez's account, which would assimilate the notion with formal falsity.[107]

Reconciliations and Condemnations

After the publication of the *Meditations*, Descartes became involved in philosophical controversies on a larger scale. He quarreled with Voetius, rector of Utrecht University, and judgment was pronounced against him by the Utrecht magistrates in 1642.[108] Perhaps because of his greater problems with the Protestants in the Netherlands, Descartes sought to make peace with the Jesuits. In 1644, after Descartes published Bourdin's *Seventh Set of Objections* and his *Replies*, together with his *Letter to Dinet*, complaining about how badly he had been treated, there was a reconciliation between Descartes and Bourdin. Descartes visited Bourdin at the Collège de Clermont, and Bourdin offered to play the role of Mersenne in Paris, to distribute Descartes's letters. Descartes also visited La Flèche itself, for the first time since he had left it. From 1644 until his death in 1650, the relations between Descartes and the Jesuits remained outwardly cordial.[109] However, in 1651 a general instruction on teaching, with lists of topics that should be excluded from university teaching, was issued after the Ninth Congregation of 1649–50 by Francesco Piccolomini.[110] The list contained several theses that were defended by Descartes. In 1663, the works of Descartes

---

sion, although there is something as it were materially false, it is not something false in the judgment that affirms or puts something forward, but merely in the sign, which signifies something false in its own right. Thus there is falsity in the proposition 'There is no God', either written down or materially put forward by him who reports 'The fool hath said in his heart "There is no God"'; and hence the account of this kind of falsity is the same as that which applies to the falsity which occurs in some verbal composition, as in a sign. But if the apprehension comes about by means of concepts of things, it is scarcely intelligible how there could be a composition of the apprehending mind without some judgment" (ACS 35–36).

107. There is, of course, a large literature on material falsity, which these few remarks cannot adequately represent; see, for example, Wilson 1978, 116–17; M. Bolton, "Confused and Obscure Ideas of Sense," in A. O. Rorty 1986, 389–404.

108. See Verbeek 1988 and 1992.

109. See, for example, AT 4:156–58, 584. In AT 4:159, Descartes tells Dinet: "Having attempted to write a philosophy, I know that your Society alone, more than any other, can make it succeed or fail."

110. "Ordinatio pro Studiis Superioribus," in Pachtler 1890, 3:77–98.

themselves were put on the *Index of Prohibited Works* with the notation *donec corrigantur*—"until corrected,"[111] and specific Cartesian doctrines were prohibited in an assembly of Jesuits and Oratorians in 1678.[112] But this did not prevent Descartes from having followers.

Descartes even picked up some followers among the Jesuits of La Flèche, though very belatedly. For example, one can find support for various early modern doctrines in a student thesis (by Ignace de Tremblay) defended in July 1700 at La Flèche.[113] One can also find a Malebranchiste and Cartesian Jesuit, the Père André, teaching at La Flèche, though not without some problems with his superiors.[114]

A final spasm of opposition to Descartes's work within the Jesuit order occurred during the first decade of the 1700s.[115] Michel-Angelo Tamburini was elected general of the order on 31 January 1706; his first act was the promulgation of thirty prohibited propositions, all of them reaffirmations of scholastic views and condemnations of Cartesian and Malebranchian positions.[116] Some of the propositions seemed to be condemnations of Malebranchian positions rather than those of Descartes. In any case, the attempt at condemnation could not have succeeded for very long; among the Jesuit propositions one even finds the denial of the relativity of motion and the denial of the conservation of inertia. Once again, however, the resiliency of Aristotelian ideas seems to have been demonstrated.

Moderns tend to think of Cartesianism as having dealt the fatal blow to Scholasticism; and this opinion, despite the surprising tenacity of Aristotelianism, has the ring of truth. However, the defeat of Aristotelianism was accomplished by tactical measures as well as by arguments and doctrines. Descartes, as we have seen, was keenly aware of this aspect of his relations with contemporaries and predecessors. In a letter to Beeckman, he wrote: "Consider first what are the things a person can learn from another; you will find that they are languages, stories, experiences, and clear and distinct demonstrations, such as those of the geometers, that bring conviction to the mind. As for the opinions and maxims of the philosophers, merely to repeat them is not to teach them.... Who teaches me, that is, who teaches anyone who loves wisdom? No doubt it is the person who can

---

111. The likely reason Descartes was put on the Index was, ironically, his attempt to dabble in theology, his account of transubstantiation. See Armogathe 1977; see also Chapter 7.
112. See Chapters 7–10.
113. Rochemonteix 1889, 4:357–64.
114. Rochemonteix 1889, 4:82–88, 94–98.
115. Cartesianism seems also to have been frowned upon by the civil authorities until 1715; see Brockliss 1992, 353. See also Chapters 8–10.
116. Rochemonteix 1889, 4:89n–93n. The text, in English translation, may be found in ACS 258–60.

first persuade someone with his reasons, or at least by his authority."[117] Descartes won some early battles by seeming to defy authority and lost others when trying to identify himself with conventional authorities; many years later, after his death, he finally won the war, perhaps by persuading others with his reasons.

117. AT 1:156.

PART I

Context

[2]

Descartes and the Scotists

To date, the most substantial works on the intellectual relations between Descartes and his predecessors have been Etienne Gilson's masterful studies.[1] In the *Index scolastico-cartésien*, Gilson catalogued various concepts in Descartes and matching ones in his scholastic predecessors. Gilson's choice of antecedents was careful. He compared Descartes's works with those of Thomas Aquinas, the Jesuits of the University of Coimbra, Francisco Suarez, Franciscus Toletus, Antonius Rubius, and Eustachius a Sancto Paulo.[2] As Gilson indicated in his introduction to the *Index scolastico-cartésien*, the teaching at Descartes's Jesuit college, La Flèche, was based on Saint Thomas, and Descartes continued to consult Thomas throughout his life. Further, Descartes became acquainted at La Flèche with the works of the Coimbran Jesuits, Toletus, and Rubius. Gilson defended the choice of Suarez by indicating that Descartes was familiar with his work—that Suarez's *Disputationes metaphysicae* was the handbook in metaphysics for Descartes's teachers. Gilson added passages from Eustachius a Sancto Paulo to those of Thomas and the others—all Jesuits—because Eustachius's *Summa*, which Descartes had read, could be said to summarize scholastic teaching faithfully and concisely.[3] It would be difficult to disagree with any of Gilson's reasons.

Descartes clearly knew the works of Saint Thomas; he even brought one of Thomas's volumes along with him on his travels—at least, that is what he said in the letter of 25 December 1639 to Marin Mersenne: "I am not at all

---

1. Gilson 1913, 1989 [1913], 1925, and 1930.
2. Gilson also adds an appendix to the *Lexicon* of Etienne Chauvin (1692). See the Appendix for an index of Gilson's *Index*.
3. Gilson 1913, iv–v.

so deprived of books as you think, and I have here still a *Summa* of Saint Thomas and a Bible that I have brought with me from France."[4] As is well known, Descartes disputed with Caterus about Thomas's views in his *Replies* to the *First Objections*, and on various occasions compared his own views on the Eucharist with those of Thomas.[5] Similarly, in a notorious passage about material falsity in the *Fourth Set of Replies* to Arnauld, Descartes cited Suarez's *Disputationes metaphysicae*. Thus Descartes must have consulted Suarez's works at least once.[6] Moreover, as we have previously indicated, the second half of the sixteenth century saw a revival of the Thomistic interpretation of Aristotelian philosophy. In 1567 Pope Pius V proclaimed Saint Thomas Doctor of the Church; Saint Ignatius of Loyola, founder of the Jesuits, advised the Jesuits to follow the doctrines of Saint Thomas in theology and those of Aristotle in philosophy. This advice was made formal in the Jesuits' *ratio studiorum* of 1586. The result was the well-known Jesuit penchant for Thomistic doctrines, at least for the period of the generalship of Claudio Acquaviva (1581–1616), when Descartes was a student at La Flèche (roughly 1607–15).[7]

As for the question of which textbooks Descartes consulted while at La Flèche, that can easily be settled. Descartes did refer to the Jesuits of Coimbra, Toletus, and Rubius as authors of textbooks he remembers from his youth.[8] As can be expected, those textbooks were often paraphrases of Thomas's reformulations of Aristotle. But seventeenth-century Thomism was not identical to thirteenth-century Thomism. At the very least, seventeenth-century Thomism was taught against the background of competing fourteenth- to sixteenth-century doctrines; and since Thomism did not have a doctrine for every possible topic, certain other doctrines and terms were imported to fill gaps. Descartes also indicated that he had read the *Summa philosophica quadripartita* of Eustachius a Sancto Paulo.[9] The textbooks mentioned by Descartes were very widely read Latin-language philos-

---

4. AT 2:630.
5. In the letter to Charlet of 31 December 1640 there is even a specific reference to Thomas's views on the Eucharist (AT 3:274). For more on these topics, see Chapter 7.
6. AT 7:235. There is a story that Descartes carried the *Disputationes metaphysicae* with him in his travels. The story is repeated by Heilbron (1979, 108), but it is unlikely at best, and, as far as I know, there is no evidence for it in the eleven volumes of *Oeuvres de Descartes* (AT). Heilbron gives three references, but they basically cite one another and lead nowhere.
7. See Chapter 1.
8. AT 3:185. For more details on the Coimbrans, Toletus, and Rubius, see Chapter 1.
9. AT 3:232. It is unlikely that Descartes had read Eustachius's *Summa* at la Flèche. Eustachius also published a *Summa theologiae tripartita* (Eustachius a Sancto Paulo 1613–16). As Descartes continued his reading of scholastic textbooks, he looked at the textbook of someone he called Draconis (AT 3:234), that is, Abra de Raconis. For more on both these assertions, see Chapter 1.

ophy texts from the first half of the seventeenth century, with Eustachius a Sancto Paulo's *Summa* probably taking first rank.[10]

Dalbiez's Critique of Gilson

With the *Index scolastico-cartésien* as his instrument of research, Gilson proceeded to work on his commentary on Descartes's *Discours de la méthode*. Immediately, a criticism of the commentary was issued by Roland Dalbiez. At stake was Gilson's comment that "in scholastic thought, objective being is not a real being, but a rational being; it does not need a special cause. In Cartesianism, objective being is a lesser being than the actual being of the thing; however, it is a real being and, as a consequence, it requires a cause of its existence."[11] Dalbiez agreed with the comment as long as "scholastic thought" was restricted here to Thomism. He pointed out that the Thomist doctrine was disputed by many previous thinkers, and especially by Scotists. For Thomists, objective being is only a being of reason; for Scotists, it is more than a being of reason.[12] Dalbiez proceeded to show that Tommas Cajetan, Thomas Aquinas's sixteenth-century commentator, dealt in some detail with Scotus's doctrine of objective reality, contrasting it with that of Thomas. And Cajetan's commentaries were published with the master edition of the works of Saint Thomas in Rome in 1570–71 (at the behest of Pope Pius V)—the edition that Descartes probably consulted.

Dalbiez also discussed an issue, seemingly tangential to the debate between Thomists and Scotists about objective being, that was being waged by

---

10. The seventeenth century also saw an enormous growth of French-language philosophy texts written by the tutors of the nobility. The movement began in the 1560s with the first French translations of Aristotle's works, but took off in the 1590s with the first French-language commentaries on the *Physics*. Works in this genre not mentioned by Descartes included Bouju 1614 (by Théophraste Bouju, an almoner of Henry IV), Ceriziers 1643 (by the Jesuit René de Ceriziers, who became a secular almoner of the Duc d'Orléans and later counselor to Louis XIV), and Marandé 1642, among others. The most noted of such works was Dupleix 1627, by Cardinal Richelieu's favorite historian, Scipion Dupleix. It exceeded Eustachius's *Summa* in popularity; more than twenty-four editions of Dupleix's *Physique* appeared in various incarnations during the first half of the seventeenth century (see Ariew 1998). Of course, there were other textbook authors with followings, who not mentioned by Descartes. For example, Henry Alsted seems to have been read by Leibniz, Rudolph Arriaga by Bayle, Franco Burgersdijk by Locke, John Magirus by Newton, and Daniel Sennert by Boyle and Leibniz. In Paris, Latin-language textbook authors included the Protestant Pierre du Moulin, and Catholics such as François Le Rees, J.-C. Frey, Jacques du Chevreul, and Jean Crassot. See C. H. Lohr's series of articles, and Schmitt and Skinner 1987. For other sources for the textbook tradition in the seventeenth century, see Reif 1962, 1969; Brockliss 1981, 1987.

11. Dalbiez 1929, 464, citing Gilson 1925, 321.

12. Dalbiez 1929, 465.

two of the Jesuit order's greatest metaphysicians, Suarez and Gabriel Vasquez. It concerned the distinction between the formal and objective concept and the role of the latter in the definition of truth. Durandus a Sancto Porciano had maintained that truth consists in the conformity of the objective concept and the thing.[13] Suarez reported and criticized this thesis: "The first proposition [of Durandus] is that truth does not reside in the formal act or cognition of the intellect, but in the thing cognized as objective in the intellect, so that the thing is in conformity with itself in respect to the existent thing, and in this way he explains that truth is the conformity of the intellect to the thing, that is, the conformity of the objective concept of the enunciative intellect to the thing according to its real being."[14] As Dalbiez said, for most Scholastics the objective concept is the thing itself, insofar as it is cognized, but for Durandus the objective concept seems to become a third reality (*tertium quid*) between the formal concept and the thing. Suarez rejected the thesis, but it was defended by Vasquez, specifically referring to Durandus.[15] Dalbiez concluded that "Descartes could not have been completely unaware of the debate. Whether Descartes' professor of philosophy was a follower of Suarez or Vasquez, he could not have neglected the exposition of a controversy that divided the two most noted doctors of the Society."[16]

The scholastic debate therefore revolved about whether the objective concept collapses into the formal concept, that is, into an act of the intellect, in which case objective being is a being of reason, or whether the objective concept is a third reality between the formal concept and the thing, in which case objective being is something more than a being of reason. It is noteworthy that this debate can also be found at the University of Paris in the first few decades of the seventeenth century, for example, in the *Summa philosophia* of Abra de Raconis.[17] De Raconis distinguishes between objective and subjective concepts: "Something can be in the intellect in two ways, either objectively or subjectively: objectively, such as man, insofar as it is the object of the intellect, subjectively, such as an intellection, that which is received in the intellect itself; as a result, there is a difficulty between philosophers, namely, in what consists the causes of exemplar ideas, whether they consist in the objective concept, in which these things are objectively in the intellect, or in the formal concept, in which these things are subjectively in

---

13. Durandus a Sancto Porciano was an early fourteenth-century Dominican who held various anti-Thomist views in physics and metaphysics.
14. Dalbiez 1929, 468, quoting Suarez discussing Durandus a Sancto Porciano.
15. Dalbiez 1929, 469.
16. Dalbiez 1929, 470.
17. For more on de Raconis, see Chapter 1.

the intellect."[18] And de Raconis asserts that there are two opposite opinions about the *ratio* of the exemplar or idea. The first is held by Thomas and Suarez: "Saint Thomas, part I quest. 15 art. 1, and, supporting him, Suarez, M 1 d 29 s2 n. 10 *et seq.*, hold that the *ratio* of exemplar causes consists in the formal concept and not in the objective"; but de Raconis thinks that the second opinion, held by Durandus, is more probable: "the essential *ratio* of the exemplar does not consist in the formal concept: Durandus, 1 Sent. dist. 36 and elsewhere."[19] He clearly favors the latter interpretation, siding with Durandus, and he supports it with various propositions.[20]

Eustachius a Sancto Paulo holds the same view as Abra de Raconis, although it is often difficult to make out exactly where Eustachius stands on particular issues, since he does not usually cite authorities or impart many details. With de Raconis's discussion as background, one can see more clearly what Eustachius intends. For Eustachius an idea is an image of a thing in the mind of the artificer; it is also an act of the mind.[21] Thus, for Eustachius, ideas or concepts can be taken in two ways, objectively or formally:

> The concept of any given thing may be taken in two senses, one formal and the other objective. The latter strictly speaking is called a concept only in an analogical and nominal sense; for it is not truly a concept, but rather a thing conceived, or an object of conception. A formal concept, however, is the actual likeness of the thing which is understood by the intellect, produced in order to represent the thing. For example, when the intellect perceives human nature, the actual likeness, which it produces in respect of human nature, is the formal concept of the nature in question, as understood by the intellect. We say actual likeness to distinguish it from the intelligible species, which is the habitual image of the same thing. It may be understood from this that the formal concept is the word which the mind possesses, or the species which it forms of the thing that is understood. The objective concept, however, which is then said to be the formal *ratio,* is the thing as represented to the intellect by means of the formal concept; thus in the example just given, human nature, as it is actually apprehended, is called the objective concept.[22]

---

18. Raconis 1651, Physica, Tractatus de causis, art. secundus, de causa exemplaris, pp. 94–95. At the risk of multiplying the terminology, in Cartesian language something can be in the intellect objectively or formally (the latter is the same as subjectively for de Raconis).
19. Raconis 1651, loc. cit., p. 95.
20. Raconis 1651, loc. cit., pp. 95–96; see also Raconis 1651, Metaphysica, tract. 3, pp. 57–68. For more details, see Chapter 3.
21. Eustachius 1629, Physica, pars 3, disp. 1, quaest. 3, p. 36. See Chapter 3.
22. Eustachius 1629, Metaphysica, pars 1, tract. 1, disp. 1, quaest. 2, p. 6; trans. in ACS 93, slightly modified.

And in case it might be thought that the objective concept would collapse into the formal concept or the thing, Eustachius specifies that

> To understand what is meant by objective being in the intellect, one must note the distinction between objective and subjective being in the intellect. To be objectively in the intellect is nothing else than to be actually present as an object to the knowing intellect, whether what is present as an object of knowledge has true being within or outside the intellect, or not. To be subjectively in the intellect is to be in it as in a subject, as dispositions and intellectual acts are understood to be in it. But since those things which are in the intellect subjectively can be known by the intellect, it can happen that the same thing can at the same time be both objectively and subjectively in the intellect. Other things which really exist outside the intellect, though they are not subjectively in the intellect can be in it objectively, as we have noted. But since all these things are real, they have some real being in themselves apart from the objective being in the intellect. There are certain items which have no other being apart from objective being, or being known by the intellect: these are called entities of reason.[23]

The doctrines of Eustachius and de Raconis are plainly an amalgam of those of Scotus and Durandus.[24] Although Durandus's doctrine, that truth is the agreement of the objective concept and the thing, might be held separately from Scotus's doctrine, that objective being is greater than a being of reason and less than a real being, the doctrines clearly complement each other. Durandus's doctrine makes sense within the general framework of a theory of distinctions that would allow for a third kind of distinction beyond real distinctions and distinction of reasons. To play its role, the objective concept must be distinguished both from the thing and from the formal concept, and those distinctions must be other than distinctions of reason. But formal and objective concepts are both in the intellect; thus, they must be distinguished formally or modally. And of course the formal distinction (or the third kind in addition to real and rational distinctions) is a notorious Scotist doctrine,[25] and one that Eustachius and de Raconis accepted. Eustachius argues that there are three kinds of distinctions: *realis*, *a natura rei*, and *rationis;* he further subdivides *a natura rei* into *formalis*,

---

23. Eustachius 1629, Metaphysica, pars 1, tract. 1, disp. 2, quaest. 3, pp. 10–11; trans. in ACS 93–94.

24. For an analysis of the relation between the doctrines of seventeenth-century Scholastics (Eustachius, de Raconis and others) and Descartes on ideas, see Chapter 3. See also Leslie Armour's discussion of Eustachius on (transcendental) truth consisting in the conformity of things to God's intellect (Armour 1993).

25. See, for example, Wolters 1990, 27–42. For a discussion of scholastic and Cartesian theory of distinctions, see Alessandro Ghisalberti, "La dottrina delle distinzioni nei Principia: traditione e innovatione," in Armogathe and Belgioioiso 1996, 179–201.

modalis, and *potentialis.*²⁶ De Raconis similarly argues for real, formal and modal, and rational distinctions in the context of debates, both real and terminological, between Thomas and Scotus.²⁷

As one can see, Dalbiez's critique of Gilson can be both extended (to other writers) and generalized (to other topics). Not only did Gilson miss the Scotism in the seventeenth-century scholastic theory of ideas—not only did he fail to notice that Scotism survives in Cajetan's commentaries on Thomas and in debates within the Jesuit order itself—but, in addition, he did not recognize that Scotist doctrines also survive in the teaching of University of Paris professors such as Eustachius and de Raconis. In fact, it can be shown that the philosophical climate in France from the early 1600s (with the major exception of Jesuit philosophy in the first half of the seventeenth century)²⁸ was predominantly Scotist and not Thomist.

What Is a Scotist?

There are, of course, no necessary or sufficient conditions for being a Scotist (or a Thomist, or even an Aristotelian). However, there are a number of different issues on which Scotus disagreed with Thomas, both major and minor, in both philosophy (including logic, metaphysics, physics, and ethics) and theology. Many of Scotus's followers took up these issues, continuing the disagreement. In the seventeenth century those oppositions were considered significant enough that some authors wrote books detailing the "two great systems of philosophy," Thomism and Scotism;²⁹ others tried to reconcile them;³⁰ still others wrote books following Thomas³¹ or following Scotus.³² Thus, the categories "Scotist" and "Thomist" are not anachronisms or historians' constructions but come from the sixteenth- and seventeenth-century writers themselves.³³ Here is a small sample of some sharp dichotomies from the theological-metaphysical-cosmological side of the curriculum (apart from the constellation dealing with the formal distinction, objective being, and objective concept that we have already mentioned).

---

26. Eustachius 1629, Metaphysica, pars 3, disp. 3, quaest. 5–8, pp. 52–55. Interestingly, Gilson cites Eustachius on the various distinctions in the *Index,* but compares him only with Suarez. See the Appendix under "distinction."
27. Raconis 1651, Metaphysica, Brevis Appendix, quaest. 1, art. 1, pp. 81–84.
28. Though again with the proviso that seventeenth-century Thomism might itself contain a fair number of Scotist terms and doctrines.
29. Or the three great systems—Thomist, Scotist, and Nominalist (or Ockhamist); sometimes the Averroist system was added. For example: Amici 1626, Rada 1620, Vincent 1660–71. See also Di Vona 1994.
30. For example, Boccafuoco 1589; Sarnanus 1590.
31. For example: Goudin 1864; John of St. Thomas 1663.
32. For example: Frassen 1668, 1677; Llamazares [1669?]; Poncius 1672.
33. In now-fashionable terminology, they are "actors' categories."

Thomas	Scotus
1. The proper object of the human intellect is the quiddity of material being (*quidditas rei materiali*).[34]	1*. The proper object of the human intellect is being in general (*ens in quantum est*).[35]
2. Only analogical predication holds between God and creatures.[36]	2*. The concept of being holds univocally between God and creatures.[37]
3. Man is a unity of single form (the rational soul).[38]	3*. Man is a composite of a plurality of forms (rational, sensitive, and vegetative souls).[39]
4. Prime matter is pure potency.[40]	4*. Prime matter can subsist independently of form by God's omnipotence.[41]
5. The principle of individuation is signate matter (*materia signata quantitate*).[42]	5*. The principle of individuation is a *haecceitas*, or form.[43]
6. The immobility of the universe as a whole is the frame of reference for motion.[44]	6*. Space is radically relative: there is no absolute frame of reference for motion.[45]
7. Without motion there would be no time.[46]	7*. Time is independent of motion.[47]

If I may be permitted a certain level of generality, the first couple of theses present Scotus's moderate Augustinianism, that is, his commitment to the doctrine that humans have knowledge of infinite being,[48] which leads him even to accept the ontological argument as in some fashion self-evident to us,[49] and not merely self-evident in itself, as Thomas would have it.[50] Most of the other theses demonstrate Scotus's attachment to the doctrine of God's absolute omnipotence, which leads him to reject or modify many propositions he thinks infringe too much upon that omnipotence.

Seventeenth-Century Scotism

It is therefore necessary to discuss the destiny of these propositions in the seventeenth century. We must ask whether or not they were generally supported by early modern Scholastics.

34. Aquinas 1964–76, I, quaest. 84, art. 7.
35. Scotus 1968 [1639], *Opus Oxoniense* I, dist. 3, quaest. 3.
36. Aquinas 1964–76, I, quaest. 13, art. 5.
37. Scotus 1968, *Opus Oxoniense* II, dist. 3, quaest. 2.
38. Aquinas 1964–76, I, quaest. 76, art. 3.
39. Scotus 1968, *Opus Oxoniense* IV, dist. 11, quaest. 3.
40. Aquinas 1964–76, I, quaest. 66, art. 1.
41. Scotus 1968, *Opus Oxoniense* II, dist. 12, quaest. 1.
42. Aquinas 1933, chap. 3.
43. Scotus 1968, *Opus Oxoniense* II, dist. 3, quaest. 6.
44. Aquinas 1953, IV, lectio 8.
45. Scotus 1968, *Quaestiones Quodlibetales*, quaest. 12.
46. Aquinas 1953, IV, lectio 16–17.
47. Scotus 1968, *Quaestiones Quodlibetales*, quaest. 11.
48. Though also attempting to avoid the extreme Augustinianism of Henry of Ghent. See Gilson's discussion of this issue in Gilson 1952, 116–215.
49. Scotus 1968, *Opus Oxoniense* I, dist. 2, quaest. 1, and elsewhere.
50. Cf. Aquinas 1964–76, I, quaest. 2, art. 1.

On the key question of whether the proper object of the human intellect, as studied by the science of metaphysics, is the quiddity of material being (with the intellect proceeding up the hierarchy of beings ultimately by analogy alone) or whether it is being in general, Eustachius sides with Scotus (proposition 1*):

> Philosophers differ on this matter. Some maintain that the object of metaphysics is God, others that it is separate substances, others that it is substance in general, others that it is finite (or so called predicated) being. All these definitions are too narrow, as will appear. Others extend its scope too far, when they say that the object of metaphysics is being taken in the broadest sense, to include both real entities and entities of reason; yet a true and real science, especially the foremost and queen of all the sciences, does not consider such tenuous entities in themselves, only accidentally. So the standard view is far more plausible, namely that the complete object of metaphysics in itself (for our question is not about its partial or incidental object) is real being, complete and in itself, common to God and created things.[51]

Without referring to any particular authority, Eustachius rejects the Thomist position that the object of metaphysics is predicated being. Interestingly, after rejecting another position as too daring, he accepts the Scotist one, that the object of metaphysics is being, common to God and created things, as the standard view. Eustachius also accepts the proposition that God's essence cannot be conceived except as existing:

> Existence belongs to God and to created things, but with a difference. For God exists not through existence being added to his nature, but through his very essence (just as quantity is said to be extended through itself). But this is not true of created things, since their existence is accidental to their essence. Hence existence is essential to God, so that it is a contradiction that he should not exist, but existence is not essential to created things, which can either exist or not exist. Hence the divine nature cannot be conceived except as actually existing; for if it were conceived as not actually existing, there would be something missing in its perfection, which is quite inconsistent with its actual infinity. But the formal or essential concept of a created thing is distinct from its existence.[52]

And, consistently with the two previous passages, he argues that we can form concepts of God's essence in this life (proposition 2*):

> By means of the natural light we can even in this life have imperfect awareness of God, not merely of his existence but even of his essence. For by the power

---

51. Eustachius 1629, Metaphysica, Praef. quaest, 2, p. 1; trans. in ACS 92.
52. Eustachius 1629, Metaphysica, pars 2, disp. 2, quaest. 4, p. 24; trans. in ACS 95–96.

of natural inference we can infer that God is an infinite being, a substance that is uncreated, purest actuality, an absolutely primary cause, supremely good, most high and incomprehensible. All these things belong to God by his very essence, and indeed uniquely, since they cannot belong to any other being. Hence when I grasp in my mind an infinite or uncreated being, or some such, I fashion for myself a concept uniquely applicable to God, in virtue of which I have imperfect awareness of his essence. Hence we can in this life form concepts of God which are unique and proper to him.[53]

These passages suggest that Eustachius was structurally or fundamentally Scotist, not Thomist. It would not be difficult to document his support of any of the other Scotist theses—to show that he also accepts the plurality of forms (proposition 3*), for example.[54]

Yet, by themselves, the passages do not show that the intellectual context of early seventeenth-century France was Scotist. For that, one has to compare Eustachius and other French Scholastics in the seventeenth century with respect to the key doctrines listed above. I discuss the issues of matter and form in seventeenth-century Scholasticism (propositions 4* and 5*) in Chapter 4. It should suffice here to assert that Eustachius thinks that prime matter can subsist independently of form by God's omnipotence (and so do de Raconis, Dupleix, and others, but not Toletus and the Coimbrans) and that he thinks that the principle of individuation is not signate matter but a form (the same for de Raconis, Dupleix, and others). What follows is a sample of such comparisons for some issues dealing with space (propositions 6 and 6*) and time (7 and 7*). Again the pattern is that Jesuits (and Dominicans) generally take Thomas's side, at least in the first half of the century, and the University of Paris doctors (and Franciscans) usually take Scotus's side.[55]

The Relativity of Space and Motion

To comprehend the debates about space one has to understand the context in which these debates were conducted, that is, the Aristotelian theory of place, which was itself developed against the backdrop of Platonic and

---

53. Eustachius 1629, Metaphysica, pars 4, disp. 3, quaest. 1, p. 71; trans. in ACS 96. Eustachius continues, however, by denying that we can demonstrate God's existence a priori, since God is not known to us *per se nota* (quaest. 2, pp. 73–74).

54. Eustachius 1629, Physica, pars 3, tract. 1, disp. 1, quaest. 6, pp. 174–75. It should be noted that Jesuits were specifically required to teach that "there are not several souls in man, intellective, sensitive and vegetative souls, and neither are there two kinds of souls in animals, sensitive and vegetative souls, according Aristotle and the true philosophy"; see Chapter 1. See also the discussion of what Emily Michael calls "Latin pluralism" in Michael 1997.

55. A more complete comparison of all of these issues has to await the completion of the monograph I have been working on, *All of Philosophy in Four Parts: School Doctrines in Descartes's Century*.

atomist conceptions of space. Plato in the *Timaeus* held that space is an everlasting receptacle that provides a situation for all things that come into being.[56] It is not clear whether Plato's talk of space as a receptacle entailed its independent existence; according to Aristotle, Plato considered matter and space the same and identified space and place.[57] Aristotle agreed. His primary concept was "place," or location in space, as one might say, space being the aggregate of all places. He defined place as the boundary of a containing body in contact with a contained body that can undergo locomotion. But he also asserted that place is the innermost *motionless* boundary of what contains. Thus, the place of a ship in a river is not defined by the flowing waters but by the whole river, because the river is motionless as a whole. These definitions gave rise to questions about whether place is itself mobile or immobile. They also engendered a problem about the place of the ultimate containing body, the ultimate sphere of a universe constituted from a finite number of homocentric spheres. If having a place depends on being contained, the ultimate sphere will not have a place since there is no body outside it to contain it. But the ultimate sphere, or heaven, needs to have a place because it rotates, and motion involves change of place. Aristotle recognized these difficulties.[58] In part, his solution was a distinction between place *per se* and place *per accidens*. Place *per se* is the place that bodies capable of locomotion or growth must possess. Place *per accidens* is the place that some things possess indirectly, "through things conjoined with them, as the soul and the heaven. Heaven is, in a way, in place, for all its parts are; for on the orb, one part contains another."[59]

Aquinas accepted and modified slightly Aristotle's account of the place of the ultimate sphere; according to him, the parts of the ultimate sphere are not actually in place, but the ultimate sphere is in a place accidentally because of its parts, which are themselves potentially in place.[60] He also rejected Averroës' popular solution to the same problem, that the ultimate sphere is lodged because of its center, which is fixed.[61] The technical vocabulary developed to interpret Aquinas's view was a distinction between material place and formal place (where, in Aquinas's vocabulary, formal place is the real ground or *ratio* of place). Place is then movable accidentally (as material place) and immovable *per se* (as formal place, defined as the place of a body with respect to the universe as a whole). Thus the ship

---

56. Timaeus 52a–52d.
57. *Physics*, 209b 11–16.
58. *Physics* 212b 7–12.
59. Aristotle 1910–52, *Physics* IV, chap. 5 (212b 12–14).
60. Aquinas 1953, IV, lectio 7; and Aquinas 1954, Opusc. LII. See also Duhem 1985, chaps. 4–6.
61. Aquinas 1953, IV, lectio 8; and Aquinas 1954, Opusc. LII.

is formally immobile (with respect to the universe as a whole) when the waters flow around it. We can note that Averroës' view required the immobility of the earth and that Aquinas's view did not, though it did require the immobility of the universe as a whole. However, the Thomist views were not universally accepted, in part because they required the immobility of the universe. This view conflicted with part of the 1277 condemnation of the complex proposition that God could not move the heavens in a straight line, the reason being that he would then leave a vacuum.[62]

Scotus and Scotists considerably modified Aristotle's and Aquinas's accounts. They rejected the distinction between material and formal place, arguing instead that place is a relation of the containing body with respect to the contained body. Place is then a relative attribute of these bodies. (They also made use of the term *ubi,* sometimes referred to as inner place, to denote the symmetric relation of the contained body with respect to the containing body). Since the relation changes with any change of either of the related terms—here the contained body or the containing body—the place of a body does not remain the same when the matter around it changes, even though the body in question might remain immobile. When a body is in a variable medium, the body is in one place at an instant and in another at an other instant; to capture what is meant by the immobility of place, Scotists said that these two places are distinct but *equivalent places* from the view of local motion.[63] On the question of the ultimate sphere, Scotus denied both Averroës' and Aquinas's solutions, claiming that heaven can rotate even though no body contains it and that it could rotate even if it contained no body—even if it were formed out of a single homogeneous sphere. (Scotus even denied that the Empyrean heaven could have lodged the ultimate sphere.)[64]

The seventeenth-century discussions of the two questions about the immobility of place and the place of the ultimate sphere generally followed the expected pattern. Toletus, for example, took Aquinas's side against Scotus on the question of the immobility of place.[65] So did Théophraste Bouju, who also kept some Averroist elements. Bouju asserted that place is movable *per se* in what he called *lieu de situation* and *per accidens* in what he

---

62. Proposition 66 (49) in Mandonnet 1908, 175–91.

63. Scotus 1968, *Quaestiones in librum II Sententiarum,* dist. 2, quaest. 6. See also Duhem 1985, chaps. 4–6.

64. Scotus 1968, *Quaestiones Quodlibetales,* quaest. 12.

65. Toletus 1589, IV, quaest. 5, "An locus sit immobilis," fol. 120r–121r. Cf. Grant 1976. Here is an abbreviated version of the doctrine, from du Moulin 1644, chap. 9, Du Lieu et du Vide: "Le lieu particulier est la superficie interieure du corps, qui touche prochainement le corps contenu. Ainsi la superficie interieure d'un tonneau est le lieu du vin dont il est plein. Ce lieu est mobile. Mais il y a un lieu immobile, a savoir celuy qui se considere au regard de l'univers."

called *lieu environnant:* "The earth . . . is in a *lieu environnant* and can also be said to be in a *lieu de situation* with respect to the poles of the world. But it cannot change place with respect to its totality; thus it is immobile in that respect and mobile only with respect to some parts that can be separated from the totality and moved into others. The firmament is also in a *lieu de situation* with respect to the earth, but it cannot change except with respect to its parts and not in its totality, in the fashion of the earth."[66]

Eustachius, in contrast, used Scotus's vocabulary: place and *ubi* are relations between the containing and contained bodies, and places are the same *by equivalence*.[67] So did Abra de Raconis, who even attributed the terminology to Scotus.[68] Eustachius also developed, very briefly, some odd views about the place of the ultimate sphere. The place of the outermost sphere is internal place or space and external (but imaginary) place.[69] This seems to be a seventeenth-century development, since Abra de Raconis and others held a similar doctrine. De Raconis discussed two kinds of place, external and internal, external being the surface of the concave ambient body, and internal being the space occupied by the body.[70] The ultimate

---

66. Bouju 1614, 1:458–59 (chap. 7: "Comment le ciel et la terre sont en lieu, et peuvent estre dits se mouvoir de mouvement de lieu"); see also 1:460 (chap. 9: "Que le lieu naturel est immobile").
67. Eustachius 1629, Physica, tract. 3, disp. 2, quaest. 1, "Quid sit locus," pp. 56–58.
68. Eustachius 1629, Physica IV, tract. 2, sec. 3, pp. 205–6.
69. Eustachius 1629, Physica, tract. 3, disp. 2, quaest. 2, "Quotuplex sit locus," pp. 58–59. For more on imaginary place, see Grant 1981, chaps. 6–7. There is a nice rejection of the doctrine of imaginary space in Ceriziers's *Le philosophe français* (Ceriziers 1643, Metaphysique, 86–90):

> On ne peut traitter de l'immensité de Dieu, sans toucher quelque chose: par ce mot d'imaginaires on entend un vuide infiny, qu'on feint au dela des cieux, ou l'on place cet estre tout parfait, de peur qu'il ne soit a l'estroit des vastes et larges voutes de l'empyrée. Ceux qui tiennent cette opinion s'appuyent de l'escriture et de la raison; de l'escriture qui assure que Dieu est plus haut que les cieux; de la raison qui ne peut souffrir qu'on limite une essence infinie. Qui ne voit que ce grand Vuide est un estre d'imagination, c'est à dire une chimere? Car ou ces espaces sont quelque chose, ou elles ne sont rien; si elles sont quelque chose de reel, on a tort de les nommer imaginaires; si elles ne sont rien, pourquoy on dit que Dieu est dans le rien? Mais quoy? Sa puissance peut creer un monde hors de celuy-ci, il faut donc qu'il soit dans cet espace, que ce nouveau monde occuperoit: je l'avoue, s'il y a un espace; mais s'il n'y en a point, il n'y est pas: or il n'y a point d'autre espace que celuy que nous y concevons possible: de sorte qu'on ne peut pas dire plus proprement que Dieu y est, que nous dirions qu'il est en ces hommes qui naitront dans le siecle à venir.

Ceriziers's opposition to any God-filled imaginary space is an interesting counterpoint to his (ambiguous) acceptance of the relativity of motion in his *Physics:* "Le lieu est donc la superficie du corps, qui nous entoure. D'ou il suit contre l'opinion du vulgaire, que nous pouvons changer de place sans nous remuer; et mesme que nous la changeons aussi souvent, que le vent agite l'air ou l'eau qui nous contient" (Ceriziers 1643, Physique, 91).

70. The distinction between external and internal place (or space) can also be found in Toletus and the Coimbrans; but they do not use the distinction to resolve the two standard

heaven is in place internally, or occupies a space of three dimensions,[71] given that the external place is the surface of the concave ambient body. "Imaginary place" thus became the standard answer to such questions as to where God could move the universe, if he chose to move it, and what there was before the creation of the universe, that is, before the creation of any corporeal substance. Imaginary places, however, were generally thought of not as real things, independent of body, but on the model of a privation of a measurable thing, like a shadow, given that a privation of a measurable thing can be measured.[72] Similarly, it was held that no time elapsed when time and the world began, but that an immense privation of time—an imaginary time—had preceded the creation.

As is often the case, it was Dupleix who stated the contrast most sharply. He held that place is immobile in itself, while bodies change places. He took it that Aquinas had a different opinion, interpreting Aquinas's doctrine of formal place as the view that one can imagine a distance from each place to certain parts of the world, with respect to which a given place, though changeable, may be said to be immobile. Dupleix raged against this doctrine: "But since all this consists only in useless imaginations, I am surprised that this opinion was received in several schools of philosophy; however, there are so many weak though opinionated brains who follow so closely the doctrine of certain persons that they would follow them right or wrong, and forget the golden sentence of the Philosopher: *I am a friend of Socrates, a friend of Plato, but rather more a friend of truth.* These are, I say, weak minds who resemble certain soldiers who would give such devoted service to a Lord that they would just as soon follow him to an unjust as to a just war."[73] Dupleix preferred a doctrine that he attributed to Philoponus and Averroës, that when air is blowing around a house, one says that the place of the house changes accidentally. The house is in the same place *by equivalence.*

On the subject of the place of the universe, Dupleix also rejected Aquinas's opinion, which he called completely wrong and a mistake: "Mais c'est abuser et mescompter."[74] Dupleix also held that the heavens do not change place or move locally, since they merely rotate within their own circumference. Ultimately, Gaultruche, a Jesuit, rejected the Thomist doctrine of place, including the Thomist doctrine that the universe cannot

---

problems about the mobility of place and the place of the universe. For more on internal and external place, see Grant 1981, chap. 2.

71. *Physica* IV, tract. 2, sec. 1–2, pp. 204–5.

72. In his article on the void, Eustachius further clarified his notion of imaginary space above the heavens by asserting that it is not a vacuum, properly speaking; Eustachius 1629, Physica, tract. 2, disp. 2, quaest 5, "An motus in vacuo fieri possit," p. 61. For more on imaginary space, see Leijenhorst 1996.

73. Dupleix 1990, 149–50.

74. Dupleix 1990, 153.

move as a whole.⁷⁵ As with matter and form, the debate about the concept of place was not completely settled by the second half of the seventeenth century.⁷⁶

In sum, while late Scholastics agreed in rejecting the independence of space from body, they disagreed about other important issues. Hidden within the debate between Thomists and Scotists on the question of the mobility/immobility of place and the place of the ultimate sphere were questions about the relativity of motion or reference for motion. Some thinkers supported a Thomist doctrine in which the motion of a body is referred to its place, conceived as its relation to the universe as a whole, a universe which is necessarily immobile; others supported a Scotist doctrine in which the motion of an object is referred to its place, conceived as a purely relational property of bodies.

. *The Ideality of Time*

In somewhat the same way as space, the concept of time involved questions about whether it is dependent or independent of bodies, whether it is mind dependent, and whether it has an absolute reference or is radically relative. One can find disagreement over such issues at the start of the seventeenth century. Many Aristotelians thought time dependent on bodies, but not mind dependent. Others sided with Augustine, thinking it independent of the motion of bodies.

For Aristotle, time is the "number" of motion, that is to say, time is the enumeration of motion. There cannot be any time without there being some change; we measure motion by time and time by motion. Consequently, there are as many times as there are motions and all are able to serve as the definition of time. The choice of a motion to measure time is

---

75. Gaultruche 1665, 2:331: "notabis vero 5. contra Thomistas; Perinde esse, an puncta illa distantiae sint realia, an solium fictitia et imaginaria. Nam saltem sunt virtualiter realia, quatenus idem per ea praestari potest, quod per realia formaliter. Atque hujusmodi quidem assignari possunt in iis spatiis, quae per imaginationem finguntur a nobis existere supra caelos. Prob. Quia mundus universus potest divinitus moveri sursum, aut deorsum motu recto; quo in casu mutaret locum; et consequenter mataret etiam aliquas id genus distantias." Lawrence Brockliss thinks that Gaultruche is representative of French Jesuits and argues that French Jesuits were not always strongly aligned with Thomism (Brockliss 1992, 1995b); he may be right. However, I think the evidence supports a developmental thesis. The one French Jesuit author of a textbook before 1665, Ceriziers, holds various Thomistic doctrines which Gaultruche will reject, including that no matter can be without form (though with a modification): "on ne doit pas pourtant nier que Dieu ne puisse conserver la matière sans aucune forme, puis que ce sont deux estres distinguez, qui ne dependent pas d'avantage l'un de l'autre, que l'accident de la substance, qui se voit separé d'elle dans l'eucharistie" (51–52).

76. Although Scotists such as Frassen seem to have had the best of the argument (Frassen 1686, 357) and others such as Barbay and the Jesuit Vincent opted for the middle ground (Barbay 1676, 261–72; Vincent 1660, 2:847–925), some Thomists resolutely maintained their position (Goudin 1864, 2:504–6).

not arbitrary, however. Although Aristotle thought that time has no reality outside of the motion it measures—"the before and after are attributes of movement, and time is these *qua* numerable"[77]—he did not think that time has no reality outside of the measurer of the motion.

Thomas seems to have accepted the Aristotelian doctrine that without motion there would be no time,[78] but Scotus rejected many elements of Aristotle's doctrine. Inspired by Augustine's theory of time, Scotus argued that even if all motion were to stop, time would still exist and would measure the universal rest.[79] The standard late scholastic view seems to have been that time began with the motion of the heavens and will end with it also. Toletus argued a Thomistic line that if there is no motion, there is no generation or time;[80] on the other hand, Eustachius argued for what may have been the successor to the Scotist line: time is divisible into real time and imaginary time, where imaginary time is that which we imagine precedes the creation of the world.[81] And Dupleix referred favorably to Augustine's account of time and talked of time measuring both motion and rest.[82] René de Ceriziers summed up the apparent consensus about time in seventeenth-century Scholasticism: "Aristotle claims that time is the number of motion or of its parts, insofar as they succeed one another. Now it is certain that time is a work of our mind, since we construct a separated quantity from a continuous one, naming it the number of motion, that is, of the parts that we designate in it. There are two kinds of time: internal is the duration of each thing or its permanence in being, external is the measure of this duration."[83] De Ceriziers then discussed a criticism of the argument that time measures rest and is thus not dependent on motion: "rest is to time as darkness is to light; it is even impossible to understand rest ex-

---

77. *Physics* IV, chap. 14, 223a 28–29.
78. "Quod tempus non sit motus neque sit sine motu," Aquinas 1953, IV, lectio 16; and "quod tempus sequitur motum," lectio 17.
79. Scotus 1968, *Quaestiones Quodlibetales,* quaest. 11. Questions about the relativity of time also gained theological inspiration through the condemnations of 1277, especially the condemnation of the proposition "That if the heaven stood still, fire would not burn flax because God [time?] would not exist," Proposition 79 (156) in Mandonnet 1908, 175–91. See also Duhem 1985, chaps. 7–8. Similar arguments were later propounded by anti-Aristotelians. For example, Bernardino Telesio had asserted that Aristotle was right about the constant conjunction of time and motion but misunderstood their true relation: "the fact that we always perceive them together is no reason to claim that one of them is the ground of the other, but only, what seems to be the case, that every motion occurs in its own time and that no motion can take place without time" (*De rerum natura* I, 29).
80. Toletus 1589, IV, quaest. 12, "An tempus sit numerus motus secundum prius, & posterius," fol. 142v–143v.
81. Eustachius 1629, *Physics,* tract 3, quaest. 2, "Quomodo distinguatur tempus a motu," pp. 63–64. See also Marandé 1642, 257: "Le temps imaginaire est celuy que nous figurons auparavant la creation."
82. Dupleix 1990, 299–303.
83. Ceriziers 1643, 2:100. The same distinction is made in Marandé 1641, 256.

cept by relation to motion"; but he limited the critique, saying, "there is no being composed out of what is not.... One can say that time is composed of instants or parts whose nature consists in existing by fleeing.... Time is distinguished from motion and the existence of the being only by the various relations that things have among one another."[84] As with the questions about space, however, the debate about time—whether it is mind dependent and whether it is dependent on motion—continued into the seventeenth century, with the majority supporting a Scotist line and Dominicans such as Goudin supporting the Thomist position.[85]

Descartes's Scotism

A recent commentator in an article about Descartes's scholastic background, detailing Descartes's possible knowledge of fourteenth-century philosophy, concluded that Descartes is "firmly rooted in a Scholastic tradition which is deeply in debt to Duns Scotus."[86] I can only agree. Descartes leans toward Scotism for every one of the Scotist theses, as long as they are at all relevant to his philosophy. It can be argued[87] that Descartes agrees that the proper object of the human intellect is being in general; that the concept of being may hold univocally between God and creatures;[88] that extension subsists independently of any form; that the principle of individuation is soul, that is, a form;[89] that space is radically relative;[90] and that time is independent of motion (i.e., propositions 1*, 2*, 4*, 5*, 6*, and 7*).[91] Proposition 3*, on the plurality of forms, is moot, since for Descartes there

---

84. Ceriziers 1643, 2:102–3.
85. For example, see Frassen 1686, 400–402 (quoting Scotus), and Goudin 1864, 2:483–85 (quoting Thomas).
86. Normore 1986, 240.
87. Obviously, every one of these propositions might require an extended defence.
88. Much as I would like to push Descartes into the Scotist camp on this issue, I am persuaded by Tad Schmaltz that I would be mistaken to do so: Descartes officially denies univocal predication with respect to substance in *Principles* I, art. 51 and in the letter to Clerselier of 23 April 1649. Jean-Luc Marion even talks about the disappearance of analogy, arguing for equivocal predication (Marion 1981, 23). This is a vexed issue for Cartesians in general. See Schmaltz 1998. See also Vincent Carraud, "Arnauld: From Ockhamism to Cartesianism," in Ariew and Grene 1995, 110–28. In his article, Carraud compares Arnauld's teaching in 1641 with Descartes's philosophy, especially as it concerns the object of metaphysics and the univocity of being. At the Sorbonne, in 1641, Arnauld taught both propositions 1* and 2*, though he was forced to retract 2*.
89. See Chapter 7.
90. This is a very complex issue; on the question of the relativity of motion, see Garber 1992, chap. 6; Des Chene 1995b.
91. Time is a difficult topic in Descartes's philosophy; but what is clear is that "time adds nothing to duration taken in general except a mode of thought" (*Principles* I, art. 57), and duration is a mode of thought or extension. In this way, duration is independent of motion

is only one kind of form, mind; however, given that that form informs another substance, extension, Descartes still has Scotist-type problems about the unity of man. Descartes, of course, also accepts the ontological argument and thinks that objective being requires a formal cause of its existence.

As Dalbiez stated, Descartes could have become aware of Scotist doctrines from a number of disparate sources. He could have been exposed to the commentaries of Cajetan in the sixteenth-century master edition of Thomas's works. He could have become acquainted with Scotist thought in the very commentaries from which he was taught—after all, they did generally give references to Scotus as an authority before rejecting his views. We should add that Descartes was often in Paris (a center of Scotist thought, we have argued) during the decade between his law degree at Poitier and his long-term retreat to the Netherlands. Moreover, Descartes frequented Dutch libraries in which he could have read Scotist philosophy: one can find Descartes's name in the student registers of the University of Frankener in 1629; he presumably registered at the University in order to use its facilities. Thus, there are many possible sources for any Scotism in Descartes's thought. Finally, we should emphasize that Descartes read Eustachius (and looked at de Raconis) in 1640. This is surely relevant to understanding any change in his thought before and after 1640—from the 1632 *Le Monde* to the 1644 *Principia,* let us say.[92]

I should return briefly to Gilson in conclusion. The question is, Given (what I believe is) very ample evidence of a Scotist climate in seventeenth-century philosophy, how did Gilson miss it? First, he was obviously not anticipating great differences in seventeenth-century Scholasticism; given the dominance of Thomism in Rome during the sixteenth century, he must have expected Catholic Schoolmen to follow along. He might not have been sufficiently attuned to differences within the Church (between the Roman Church and the Gallican Church, for example). Gilson thought that Eustachius faithfully summarized the teaching of the schools, but there was no reason to believe that those doctrines were the same in all schools, in the colleges of the University of Paris as well as in the Jesuit colleges (not to mention possible individual differences among philosophers).[93] It would be worthwhile to remember that there were significant cultural and political differences between scholars and teachers of the University of Paris and those of the Jesuit colleges. Throughout the sixteenth

---

(although there is a dispute about whether motion is independent of duration). Cf. Garber 1992, 173–76.

92. See Chapter 1.

93. For an example of an author influenced by Gilson's treatment of Eustachius as a Thomist, see Carriero 1990, especially the methodological remarks in the preface, pp. v–ix.

and seventeenth centuries the University of Paris feuded with the Jesuits; the University waged three separate legal battles to prevent them from establishing a college in the city. From the 1560s, when Collège de Clermont (the Jesuits' main college in Paris) was first established, to their expulsion from France in 1595, to their reestablishment in France (and the subsequent establishment of the Collège de La Flèche in 1604), to the reestablishment of Collège de Clermont in 1616—even as late as the mid-1640s—the University tried to stop the Jesuits from teaching in Paris; it even refused to recognize degrees bestowed by them as equivalent to its own degrees for admittance into its graduate faculties of Law, Medicine, and Theology.[94] The cultural opposition between Paris and the Jesuits was just as keen as their legal battles. As J.-R. Armogathe reminds us, Jesuits were foreigners: "in the France of Henry IV, 'modernity' spoke Latin; the use of French was the domain of the 'bons français,' who were convinced that what came from beyond the Alps—Jesuits, Italians, Latin—could only ruin the purity of the ancient ways that they wished to preserve or rediscover."[95]

Naturally, political and cultural differences by themselves do not entail directly doctrinal differences. The question about whether Eustachius's philosophy was the same as that of the Jesuits and Thomas might not have needed to be answered as long as Gilson limited himself to collecting various doctrines without actually making any claims about the relationship between Descartes and the Scholastics as a whole.

Gilson's expectations might have been reinforced by the resurgence and dominance of Thomism for Catholics in the late nineteenth and early twentieth centuries (due to the 1878 encyclical *Aeterni Patris* of Leo XIII, a kindred spirit to Pius V). In any case, not expecting great differences among seventeenth-century Scholastics, Gilson did not try to locate any. In the *Index*, Gilson did not generally quote Eustachius and Thomas and the Jesuits on the same topics (see the Appendix). And when he did, it was usually on different aspects of the issue. Given the passages Gilson does quote in the *Index*, it would be difficult to determine that the philosophers cited held conflicting opinions (even if one can sometimes sense their divergences). Had he probed deeper, Gilson would have found that he was right in saying that Eustachius "faithfully summarizes scholastic teaching," but not in the way he meant it. Ironically, Eustachius did faithfully summarize the teaching of the University of Paris, a teaching that leaned toward Scotism rather than Thomism.

---

94. See Douarche 1970.
95. J.-R. Armogathe, "L'approche lexique en histoire de la philosophie," in Fattori 1997, 59.

[3]

Ideas, in and before Descartes

with Marjorie Grene

What John Locke called "the new way of ideas" governed philosophy for more than a century, years that were both fruitful and fateful in the history of philosophy. So central were "ideas" to the philosophy of this period that Antoine Arnauld and Pierre Nicole could write at the head of the first chapter of their widely adopted *La logique, ou l'art de penser:* "Some words are so clear that they cannot be explained by others, for none are more clear or more simple. 'Idea' is such a word. All that can be done to avoid mistakes in using such a word is to indicate the incorrect interpretations of which it is susceptible."[1] Now, the *Port Royal Logic,* as it is called, was more than a "logic"; it was the first standard text of modern philosophical method. What was the concept intended by this "clear" and "simple" word? And in particular, we want to ask, what was the historical context in which this seemingly perspicuous term took hold? In other words, where did the "new way of ideas" come from?

Traditionally, the term in its authoritative modern sense is attributed to Descartes. Thus, for example, L. J. Beck wrote of Descartes's usage: "It is notorious that Descartes' use of the word 'idea' is peculiar to himself in that previously the term was used to describe the Ideas of Plato and had no current usage in the terminology of the Schools."[2] And Descartes himself gives a similar impression when he tells Hobbes: "I used the word 'idea' be-

---

1. Arnauld 1964, 31.
2. Beck 1965, 151. Beck states that "for once he [Descartes] also invents a new technical terminology"; he continues by asserting that "the term itself and its usage was taken up by Gassendi and spread into England with Locke." For more on "idea" in Locke and British philosophy, see R. Hall, "*Idea* in Locke's Works," and John W. Yolton, "The Term *Idea* in Seventeenth and Eighteenth-Century British Philosophy," in Fattori and Bianchi 1990, 255–64, 237–54.

cause it was the standard philosophical term used to refer to the forms of perception belonging to the divine mind, even though we recognize that God does not possess any corporeal imagination. And besides, there was not any more appropriate term at my disposal."[3] This remark certainly seems to confirm Beck's reading.[4] Descartes says he is borrowing a term used to refer to God's ideas (the post-Augustinian descendant of the Platonic or neo-Platonic "idea"), and he remarks elsewhere that he is using it more generally for "everything which is in our mind when we conceive something, no matter how we conceive it."[5]

What we want to do here is to challenge the standard reading of this situation and, a fortiori, expand the Cartesian account. We want to ask how the term "idea" was used in the seventeenth century before Descartes, and in the light of this evidence we want to consider both the possible sources of Descartes's usage and its originality.

First, then, was the term "idea" current in the seventeenth century and how was it used? Two rather different contexts must be distinguished: a novel usage in French and English literature, which is quite unequivocal, and a more obscure and ambiguous complex of philosophical definitions.

In the literary usage, "idea" refers to images, usually derived from sense. This is clear from dictionaries of the period. Jean Nicot, for example, in his 1606 *Trésor de la langue française tant ancienne que moderne*, defines ideas as follows: "*Ideas* are imaginations that people construct in their thoughts: *Ideae, idearum*. These are also the images of things that are impressed on our soul. Platonists say that there are some eternal models and portraits of all things in God, which they call ideas."[6] Modern dictionaries of this period also give the term its image-related sense. Thus Huguet's dictionary of sixteenth-century French gives image as the first meaning, citing a passage from Rabelais: "Puys nous demanda: Que vous semble de ceste imaige? — C'est (respondit Pantagruel) la ressemblance d'un pape.... Vous dictez bien (dist Homenaz). C'est l'idee de celluy Dieu de bien en terre, la venue duquel nous attendons devotement."[7] The *Dictionnaire Robert* lists the meaning of *représentation* as beginning in the seventeenth century.[8] A similar use

---

3. AT 7:181. CSM 2:127–28. We wish to restrict our discussion of idea in Descartes to this later conception; there is in Descartes an earlier conception of *corporeal ideas*. For a discussion of the earlier concept, see Michael and Michael 1989. See also Armogathe, "Sémanthèse d'*idée/idea* chez Descartes," in Fattori and Bianchi 1990, 187–205, for the different senses of idée/idea in Descartes.

4. Though not everybody accepted Beck's reading; see, for example, Urmson 1967, 4:118–21.

5. AT 3:393; CSMK 185.

6. Nicot 1979.

7. Huguet 1949, 4:536. Huguet also quotes Palissy, Aubigné, de Cornu, Montaigne, and de Sales using *idée* as image. He also gives as a second meaning "modèle, type parfait, idéal."

8. Robert 1953, 3:575.

is given in the *Oxford English Dictionary* as current from the late sixteenth century.[9] Dictionaries of earlier usage, on the other hand, such as F. Godefroy's *Dictionnaire de l'ancienne langue française et de tous ses dialectes du IXe au XVe siècle,* have no entry for "idée" whatsoever. By the end of the seventeenth century, of course, the Cartesian conception itself has entered into lexical accounts, as for example in Furetière's *Dictionnaire universel,* where the second of five meanings includes an explicit reference to Descartes's usage.[10]

It is, then, clearly in accordance with this new literary usage that Descartes calls ideas in Meditation III "as it were images of things." No wonder Hobbes took him to be following the doctrine in which ideas were identified with images. At the same time, of course, Descartes's statement to Hobbes also suggests his opposition to this equation; he used the word that people employed to designate the concepts in God's mind, although God has no corporeal imagination. Our ideas, like God's, are concepts, mental acts, or mental contents, but decidedly not images. Other passages explicitly stress this difference, as against the Hobbesian (or Gassendist) identification of idea and image. Thus, for example, in July 1641, Descartes writes to Mersenne: "by 'idea' I do not just mean the images depicted in the imagination; indeed, in so far as these images are in the corporeal imagination, I do not use that term for them at all."[11] And there follows the statement already quoted: "Instead, by the term 'idea' I mean in general everything which is in our mind when we conceive something, no matter how we conceive it."[12] Here Descartes has been discussing the comments of an unknown correspondent about his use of "idea." He continues:

> But I realize that he is not one of those who think they cannot conceive a thing when they cannot imagine it, as if this were the only way we have of thinking and conceiving. He clearly realized that this was not my opinion, and he

---

9. *Oxford English Dictionary* 1989, 3:614. The *OED* also gives "idea" its meaning of image from at least 1589. Classical Latin dictionaries do not list *idea* at all, even though the term can be found in Seneca. *Smith's Smaller Latin-English Dictionary,* a later-classical and medieval-leaning dictionary, does list *idea,* but as Platonic idea or archetype, referring to Seneca's *Epistles.*

10. Furetière 1690, II: "Idée, se dit aussi des connaissances que l'esprit acquiert par le rapport & l'assemblage de plusieurs choses qui ont passé par les sens. Descartes prouve nettement la necessité de l'existence de Dieu par l'idée qu'on se forme naturellement d'un Tout infiniement parfait, dont l'existence est une de ses perfections." See also the *Dictionnaire de l'Académie française,* for the now standard different meanings of *idée.* One also senses Descartes's influence in the database of the INaLF, which is predominantly post-Cartesian; see Gérard Gorcy, "*Idée*(s) dans le corpus textuel de l'INaLF du dix-septiéme siècle (1601–1715), Descartes et Malebranche exceptés," in Fattori and Bianchi 1990, 155–86, especially the interesting pre-Cartesian occurrences included on pp. 160, 162–63, 166, 176, and 180.

11. AT 3:392–93. CSMK 185.

12. AT 3:392–93. CSMK 185.

showed that it was not his either, since he said himself that God cannot be conceived by the imagination. But if it is not by the imagination that God is conceived, then either one conceives nothing when one speaks of God (which would be a sign of terrible blindness) or one conceives him in another manner; but whatever way we conceive him, we have the idea of him. For we cannot express anything by our words, when we understand what we are saying, without its being certain thereby that we have in us the idea of the thing which is signified by our words.[13]

Thus Descartes appears to be drawing on the current literary usage, in which ideas are not just exemplars in God's mind but actual psychological events in our minds, while at the same time refusing the identification of idea and image that the new literary sense suggests. So we must ask, further, What sources did Descartes find in the philosophical literature of his own time on which to ground his own usage? Where did the current image-oriented use appear in the philosophical as against the literary works of the period, and, on the other hand, how does the conceptual (non-image) use Descartes was to devise relate to the philosophical use of "idea" in general?

We will suggest answers to these questions by referring to a number of early seventeenth-century philosophical writers. Not that Descartes was directly influenced by one or more of them. Even though he read some of them at some time in his life, the important point is that they were well-known thinkers whose terminology would have been familiar to any scholar of the time, whether to Descartes himself or to those in his circle. The four main examples we will discuss, in chronological order, differ in their form of exposition and in the professional status of their authors. Three authors taught at the University of Paris during the first few decades of the seventeenth century: Eustachius a Sancto Paulo, Jean Crassot, and Charles François d'Abra de Raconis; the other was Rudolph Goclenius, the author of a celebrated philosophical dictionary.[14] All four will impart some clues to the revised meaning Descartes will construct.

But first, let us look at a thoroughly traditional text, which will indicate the context or occasion for the discussion of ideas in the philosophical corpus of seventeenth-century Scholasticism. The following passage can be found in the 1614 *Corps de Philosophie* of Théophraste Bouju:

> To these four kinds of causes we have just spoken of, the Platonists add a fifth, which they call exemplar or idea; for insofar as God is the universal artisan of

---

13. AT 3:393; CSMK 185. Cf. AT 3:382; 8:1, 21; 7:67, 139, 165, 179, 180, 185.
14. Goclenius has a certain usefulness, since he attempts to summarize as precisely as possible *all* the philosophical distinctions current at the start of the seventeenth century. Thus, we can use him to represent the important doctrines prior to 1613, such as Suárez's (1597), allowing us to concentrate on the milieu at Paris circa 1609–18.

all things and only makes things wisely and perfectly, understanding what he makes and why he makes it, there must be ideas, intelligible notions or forms, in his divine understanding, of the things he makes. This exemplary form is also found in the understanding of men; for in this way the natural agent has in himself the natural form by which he produces his effect and renders it similar; similarly the agent who acts through the understanding has in himself the intelligible form of what he wants to do, trying as much as possible to make what he is making resemble it. Thus the doctor tries to introduce health to his patient in accordance with the idea he has of it, and the architect to construct a house materially similar to the one in his thought.[15]

Bouju, sieur de Beaulieu, the king's almoner and counselor, is writing an ordinary philosophy textbook in French (for those not comfortable or not educated in the Latin of the schools). His account is entirely traditional.[16] The context in which he discusses ideas is the standard one. He has been enumerating the Aristotelian causes and adds a discussion of "exemplary causation." Ideas are routinely identified with exemplars, that is, either Platonic ideas or ideas in God's mind. And the question is whether, in serving as models for creation, ideas (exemplars) *cause* the things that imitate them in some fifth way. Further, the architect in building a house tries to make it like the one "he has in his mind," the physician has an idea of health, and so on. Now this sounds at first glance very like the psychological meaning of idea that we have discussed in the literary usage of the period. Granted, the physician's idea of health is perhaps not an image, but it is surely something psychological. It soon emerges, however, that Bouju is chiefly echoing a well-established scholastic-Aristotelian tradition, in which ideas are either the forms in God's mind according to which he makes things or the exemplars in artificers' minds when they make their objects, houses, statues, or paintings. Ideas as exemplars, however, are not strictly psychological, like Rabelais's or Hobbes's images. They are general forms, patterns to be followed in this or that case, and not particular mental events. Still discussing the question of causality, Bouju continues:

> But this cause is not of another kind than the four we have posited, since, according to the opinion of most philosophers, it reduces to the formal separated and external cause; as the thing is determined and derives its specific perfection from the form which is part of the composite and its internal cause,

---

15. Bouju 1614, 1:297–98 (chap. 113, "De la cause exemplaire"). The talk of exemplar causes as a Platonic fifth kind of cause to be added on to the four Aristotelian causes dates back at least to Seneca, *Epistle* no. 65.

16. The seventeenth-century Scholastics Descartes is known to have read in his youth are also fairly traditional. See, for example, the Conimbricences 1596, lib. 2, quaest. 3 and 4, and Toletus 1589, lib. 2, cap. 3, quaest. 7.

so also, in the same way the work (*effet*) of the artisan is in its way determined according to its particular perfection through the exemplar that resides in his mind, to which he refers when making the artificial thing by introducing something similar in it.[17]

Bouju's argument echoes Thomistic usage: "Ideas are exemplars existing in the divine mind, as is clear from St. Augustine. But God has the proper exemplars of all the things that He knows; and therefore He has ideas of all things known by Him."[18] This is of course the standard sense Descartes is referring to in his reply to Hobbes. Earlier in the same passage St. Thomas had extended the reference to the natural (and artifactual), as well as the divine. He begins, as usual, with the divine mind: "It is necessary to posit ideas in the divine mind. For the Greek word *Idea* is in Latin *Forma*. Hence by ideas are understood the forms of things, existing apart from the things themselves. Now the form of anything, existing apart from the thing itself, can be for one of two ends; either to be the exemplar of that of which it is called the form, or to be the principle of the knowledge of that thing, according as the forms of knowable things are said to be in him who knows them. In either case we must posit ideas."[19] Similarly, "the likeness of a house pre-exists in the mind of the builder. And this may be called the idea of the house, since the builder intends to build the house like the form conceived in his mind."[20] Here, as for Bouju, the idea that is "in the mind" is a form rather than a particular mental act. It is an analog of the patterns in God's mind, where ideas primarily exist.

That there are exemplars in God's mind raises an important side issue: whether the intellectual soul knows material things in the eternal exemplars. Aquinas cites as an objection what might be thought to be the standard scholastic position, that the soul does not know the eternal exemplars because it does not know God, in whom the eternal exemplars exist, and that the eternal exemplars are known through creatures and not the converse. He also objects that if we say that the intellectual soul knows all

---

17. Bouju 1614, 1:297–98. Scotus, in his well-known question on the existence of God, resolved the issue by asserting that exemplar causes are efficient causes: "the causality of an exemplar, which is added, is not a different kind of causality than efficient, since there would then be five types of causes. And so the exemplar cause is some kind of efficient cause, which acts through the intellect as distinct from acting through nature." Scotus 1968, *Opus Oxoniense* I, dist. II, quaest. 1 & 2; also *Opus Oxoniense*, book I, dist. 36, quaest. unica.

18. Aquinas 1964–76, I, quaest. 15, art. 3 (Aquinas 1945, 166); see also Jordan 1984 and R. Busa, "*Idea* negli scritti di Tommaso d'Aquino," in Fattori and Bianchi 1990, 63–88. Interestingly, by the start of the seventeenth century, the ideas in God's mind include ideas of things that will never be produced: "Mais Dieu estant infiny il y a en soy des idées d'une infinité de choses, lesquelles ne seront jamais idées pratiques, parce qu'il ne produira point des choses respondantes à icelles" (Dupleix 1992, 189).

19. Aquinas 1964–76, I, quaest. 15, art. 1 (Aquinas 1945, 161–62).

20. Aquinas 1945, 162.

things in the eternal exemplars (knows as "principle of knowledge," not as "object"), we return to Plato's opinion that all knowledge is derived from the exemplars (assuming, with Aquinas, following Augustine, that the Platonic forms are permanent exemplars in the mind of God). Aquinas denies these objections and asserts that the intellectual soul knows all truths in the eternal exemplars, distinguishing between the soul in its present state of life, which cannot see all things in the eternal exemplars, and the blessed, who see God and all things in him and thus know all things in the eternal exemplars.[21] Bouju follows Aquinas: "Some have wanted to say that things are true insofar as they resemble and are in conformity with the idea of their essence and nature which is in God. To this I respond that there is no doubt that true things are in conformity with ideas which are in the divine understanding; but it is not there that we must refer the proofs of human knowledge according to the reasons of philosophy, that is, to know whether something is true; otherwise, we would never have any certainty, given that it is outside our power to know these ideas during this life."[22]

Bouju gives us an instance of a standard approach to the nature of ideas with no hint of a new, psychological sense. On the other hand, the four authors we will now consider do suggest, in various ways, a revision of the traditional meaning of the term.

The first instance of the psychological usage in the philosophical literature seems to occur in the first part of the *Physics* of Eustachius de Sancto Paulo, *Physics* being the third part of his *Summa Philosophica Quadripartita*. Eustachius was educated at the University of Paris, receiving his doctorate from the Sorbonne in 1604. Like Bouju, he is also writing an ordinary philosophy textbook, but one in Latin instead of the vernacular. When Descartes needed a concise textbook in philosophy in order to review school philosophy and to compare it to his own, he made use of Eustachius's *Summa,* subsequently calling it the best book of that kind ever written.[23]

As was customary, *idea* is taken by Eustachius to be synonymous with *exemplar,* and exemplars are discussed under the topic of causation, corresponding to Aristotle's analysis in book 2 of his *Physics*. The question is whether exemplary causes constitute a fifth class in addition to the canonical four. Eustachius's answer is that in the case of natural causation exem-

---

21. Aquinas also asserts that there is a sense in which the human soul, even in its present state, knows all things in the eternal exemplars, since we know all things by participation in the exemplars: "For the intellectual light itself, which is in us, is nothing other than a participated likeness of the uncreated light in which are contained the eternal exemplars." Aquinas 1964–76, I, quaest. 84, art. 5 (Aquinas 1945, 389–92).
22. Bouju 1914, 1:177 (chap. 13, "Que la verité des choses ne nous est point connue par leur rapport aux idées qui sont en l'entendement de Dieu").
23. AT 3:231; CSMK 156.

plary cause may be taken to be a kind of efficient cause, and in the case of an artificer it belongs (more obscurely!) to formal cause. What concerns us here, however, is not this traditional question (which we shall return to in any case in connection with our other examples), but the nature of idea as the equivalent of exemplar. Eustachius writes:

> What the Greeks call Idea the Latins call Exemplar, which is nothing else but the explicit image or species of the thing to be made in the mind of the artificer. Thus the idea or exemplar is in this case some image (*phantasma*) or work of imagination (*phantasiae*) in the artificer to which the external work conforms. And so in the artificer insofar as he is an artificer there are two internal principles of operation, namely the art in his mind or reason and the idea or exemplar in his imagination (*phantasia*). Art is a certain disposition, but idea is a certain act or concept represented by the mind. So, the mind first represents a copy of the thing to be made through art, then it contemplates what it has represented, and directs the external work to its likeness.[24]

Note that idea here *is* an image—and, what is particularly crucial for the Cartesian reading, it is "an act or explicit concept of the mind." In this brief passage, then, we have the contemporary meaning that Descartes will exploit; an idea is, as it were, an image—expressive of something—something which the mind contemplates. It is both something I do—an act—and something I "see": *illam expressam intuetur*. What Descartes makes of the "image" part of the story is another question, to which we shall return. In the meantime, let us look at some other versions of the use of exemplar or idea in the same period.

The second case, the *Lexicon* of Goclenius, both echoes the tradition and illustrates the complexity of the issues involved. Goclenius's dictionary was a standard source for philosophical usage in the seventeenth century (as it still is for us). Ideas, as we have seen in our previous cases, are identified with exemplars; thus, Goclenius's dictionary should be consulted under both headings, as well as under the related heading "species," and also under *conceptus,* which, as we have already noticed, is equally important in the Cartesian case. ("Concept" also occurs in our brief citation from Eustachius.)

"Exemplar," Goclenius says, "is twofold, Created and Uncreated. Plato in the *Timaeus.* The uncreated exemplar is the idea in the divine mind. The created exemplar is the universal species caused by the thing. Scotus distinct. 3, bk. l. Thus there is a double conformity and a double truth, one

---

24. Eustachius a Sancto Paulo 1609, Physica, pars 3, disp. 1, quaest. 3, "Quid sit exemplar, & ad quod genus cause revocandum sit," p. 36.

conformity to the Created exemplar, the other to the Uncreated."[25] The first, uncreated exemplar, of course, is exactly what Descartes refers to as his precedent for the use of "idea." Note also that truth consists in the conformity of the thing to the exemplar/idea. Although the truth of clear and distinct ideas, for Descartes, will run in the other direction, from ideas to things (a point we will return to later), it is worth noticing that the close relation Descartes will establish between ideas and truth does seem to find a precedent here.[26]

The entry for "idea" is long and complex, including a relatively detailed historical section about Plato and Platonic commentators. Two main entries are given, though not really about separate senses of the term:

1. [Idea] signifies the species or form, or external reason (*ratio*) of the thing (which is outside the thing). Aug. l. 83. qq. quaest. 46. And thus distinct from the thing. Nor is it the form of the thing, but one of the four causes [note the causal context, as with both Eustachius and Bouju]. Its description is general or special. In general the idea is the form or exemplar of the thing, which the maker is contemplating when he makes that which he is aiming at in his mind. Seneca Ep. 66, as the painter has in mind the exemplar of that image which he can or wishes to paint.[27]

At first reading this important passage may sound like the literary usage: the idea would be the image in the painter's mind ("ideas are as it were images of things"). However, what the painter has in mind is the *exemplar* of the image: that is, the species or form—the *eidos* (the Greek equivalent of species, identified as such by Goclenius under that heading).[28] And the "image" here is of course the painting—nothing mental at all, but the actual object that the artist "can or wishes to paint," itself a copy of the idea, its model. Thus the idea here, far from being the particular image in the artist's imagination, is its very contrary: the image is the product, and the

---

25. Goclenius 1964, 196.
26. As it did, in a limited fashion, with Aquinas and Bouju. Eustachius's view is discussed in Armour 1993, 4–14.
27. Goclenius 1964, 208.
28. Peter Dear quotes an interesting passage from Fonseca, the main Coimbran and Aristotelian commentator in his own right, about the indifference of translating *idea* as *forma* or *species*. Fonseca remarks that "truly, dialecticians are not so much hampered by a religion of words" (*Institutiones*, p. 54, as quoted in Dear 1988, 21). Fonseca is obviously a good source for discussions of such terms. In fact, a recent article by Norman Wells (1993) treats Descartes in the context of Fonseca and Suárez. In most respects, this article is complementary to our view in this chapter, except insofar as Wells claims that Descartes thinks of idea as a form, citing his reply to Arnauld (AT 7:232). However, Wells's evidence is not conclusive; in *Replies* IV Descartes seems to be adopting what he thinks would be Arnauld's vocabulary and drawing out its consequences, not committing himself to ideas being some kind of form.

idea is the object to be imitated in that productive process. More surprisingly, in Seneca's account—which will be echoed, though not referred to, in our next example—it is the actual, real-life model that *is* the *idea* on which the painter bases his image, in this case, his portrait. Thus in Seneca's account, the artist is painting Virgil, and Virgil himself, the poet in the flesh, is the model = exemplar = idea. Thus Seneca's discussion of ideas, which seems to have been a standard source,[29] entails a dramatic reversal of the original Platonic conception. Rather than serving as otherworldly models for their real-world imitators, "ideas" become real-world models for the artist's use. At the same time, every account of ideas does retain enough of Plato's conception (exemplified in the *Timaeus*) to make ideas in God's mind the models of creation, themselves the primary reality, in imitation of which particular things are made. What Seneca's account, and here Goclenius's, seem to be drawing on, then, is the analogy between ideas in God's mind, the exemplars for his creation, and forms in the created world, the exemplars that the artist's work imitates. God makes things that imitate His ideas, and the artist in turn makes imitations of those created realities. "Things" are images of God's ideas, while the images made by artificers imitate *those* images. The psychological act by which the artist understands, or "sees," an image *of* his eventual work is simply omitted from this account.

So much for the first account of idea "in general." Goclenius continues: "Specifically, or in particular, [the idea is] the form, or the rationale (*ratio*) of the thing, in the divine mind, eternal and immutable, which he envisages when he makes something similar to it. This is suitable only for God. Or it is the divine essence itself as known by God, as that according to the likeness or imitation of which each thing itself is produced, and in which also the thing as in its eminent exemplar is known by God."[30]

Goclenius asserts further that the essence of God is "both the principle and the object of knowledge." He explains what he means by this: "The principle and objective knowledge of a thing is that in which the thing is known as in its likeness. For it is fitting that there be some similarity between the idea and the thing.... Thus the idea is the principle of operation in the fashion of an exemplar, since the thing is to be made in its likeness.... And the idea is in the divine or created mind."[31] All this supports the traditional conception Descartes claims to be relying on. The signal exception is that Descartes seems to have had no interest in the created as

---

29. See, in particular, Panofsky 1968.
30. Goclenius 1964, 208–9.
31. Goclenius 1964, 209.

against the uncreated context. The case of the artist does not seem to have interested him.[32]

Goclenius adds a second, briefer account of idea: "What idea is: 1. Idea is the architectural rationale (*ratio architectatrix*) in the mind of the maker, that is, the reason according to which the fabrication is carried out."[33] Note that this is, once more, not so much a particular psychological act on the part of the maker as it is a model, a norm: an exemplar to which his product is to conform. Goclenius continues:

> It is therefore a relative Being, that is, the essence of the idea is said to consist in its relation to something else, or to be referred to something else, that is, to be the exemplar of something else or its archetype. Thus both the first idea and the first exemplar are called *archetypon* exemplar.
>
> The first idea, or first exemplar, is sometimes accepted as a thing subsisting in itself, as when the idea of the World is said to be sempiternal, where the sempiternal and immutable wisdom and reason of God, by which he makes the world, is understood, that is, God himself. The Platonists call the world *noeton*, that is, intelligible, most remote from our eyes, to which they oppose the sensible world, which we inhabit.[34]

This appears overall to repeat the more or less traditional Platonic view. Moreover, it confirms Descartes's intention of separating ideas from sense. Ideas are intelligible and other than what is sensed. That may be why the notion of idea as God's concepts was a useful source for him, since God is noncorporeal and thinks without sensing. So far, again, this is very different from the identification of idea with (sensible) image that Hobbes and Gassendi favor.

Finally, let us look at Goclenius's entries under "species" and "concept." Species or *eidos*—that is, form—is given eight meanings, of which the most important for our purposes is the third: "species is taken from the image, whether conceived in the mind, 1 de anima chap. 12, or pictured or carved by the artificer, whether it is empty, or representing or expressing a true thing."[35] Here, as in Eustachius, we find, at least in part, an identification of idea with (psychological) image: "the image conceived in the mind." But the image here is *either* made or carved *or* imagined: "conceived in the mind." Thus we seem to have two meanings in one: the psychological one

---

32. Gassendi, who clearly knows the tradition, tries to get Descartes's reaction to it, referring to a painter who is the cause of the image displayed on the canvas (AT 7:289) and an architect who makes up an idea of a house in his mind (AT 7:298). This does not elicit much of a response from Descartes; see AT 7:369.
33. Goclenius 1964, 209.
34. Goclenius 1964, 209.
35. Goclenius 1964, 1068.

common in literature, where *eidos* is the artist's mental image (or conception), and the more realistic sense in which the image is what the artist produces—painting or statue—as a copy of an external entity. (It may even be empty; suggesting a place for imaginary entities as the products of the artist's work; he may paint a chimera or carve a hydra.) In any event we have here a second occurrence in the philosophical literature of the psychologized "idea" on which Descartes was to build his conception.

Additionally, attention should be drawn to Goclenius's definition of "concept" (*conceptus*). Except for the indirect link from exemplar to species, then to idea, and finally to species as *mente concepta*, there is no immediate connection here with Descartes's idea as "anything the mind can conceive." Nevertheless, Goclenius's definition is significant for Cartesian usage, especially because of his distinction between formal and objective concepts (reminiscent of Descartes's formal and objective reality applied to ideas in Meditation III). Goclenius writes: "*Concept* is not used univocally by the Scholastics. In one way Concept(ion) is, improperly, the act of conceiving or of knowing. In another way it is, *properly* the thing conceived (as object) with the act of knowing, given that it signifies the former and truly connotes the latter, just as *phantasm* signifies chiefly the thing itself seen as particular and connotes the act of imagining."[36]

Descartes's clear and distinct idea seems to conflate these two aspects; it is both the act of thinking and what is thought of *as* the object of thought. But Goclenius goes on to enumerate the two distinctions we just mentioned, which will be important for the Cartesian argument: "The formal concept is that which we form concerning something apprehended by the intellect. The objective concept is the thing which is conceived insofar as it is the object of our formal concept."[37] Here, indeed, Goclenius is setting forth the distinction much as Descartes himself will use it. Formally an idea is an act of the mind, but it also exists as content: as the object of the formal act. In Cartesian terms, an idea (or conception) has formal reality as my act of thinking and objective reality as representative of what I am thinking about. (Note, by the way, Goclenius's disagreement with Eustachius. For the former, it is the particular mental act that "idea" or "concept" properly signifies.)

Goclenius goes on to identify "conception" as synonymous with concept; the latter, he specifies, is either clear or obscure—another celebrated Cartesian distinction—but here that distinction is given a theological reference: "The Scholastics employ this distinction when they deal with the conception of God. For they call the abstractive obscure; that, moreover, is

---

36. Goclenius 1964, 427.
37. Goclenius 1964, 428. There is, of course, a sizable literature on the objective and formal being of ideas in Descartes and his predecessors, from Etienne Gilson and his critics to the present. For references to this literature, see Grene 1991; see also Chapter 2.

what is made through the species of created things or as it shines in created things: the intuitive indeed, which is as it is in itself, they say is clear."[38] Though applied to the conception of God, not to human conception, this account recalls the "intuition" of Descartes's early *Rules,* the incomplete forerunner of the *Meditations.* Clear and distinct ideas are the successors to intuitions as the units of Cartesian knowledge.[39]

In summary, we find a number of possible sources for the Cartesian idea—in the definitions of "idea" itself as well as in the related terms "species," "form," "concept," and "conception." These definitions are not yet molded into the unity Descartes will make of them, and they still contain aspects that he will all but abandon. But there is certainly a rich store of connotations here on which the (relatively) new Cartesian sense of the term could have drawn.

Our third example is from the *Physics* of Jean Crassot. Crassot was a professor of philosophy at the University of Paris; the work we are citing was published posthumously in 1618. He was well known for his lectures, which he presented in outline form, and his published text corresponds to his lecture style. In his discussion of the causes he includes a chapter on exemplars and ideas—here again exemplar and idea are linked. Exemplar, he says, "sometimes means specimen and example, sometimes archetype or something in whose likeness something is intentionally worked out."[40] Further, exemplar "is said to be example, type, Idea, *eidos,* but properly idea, exemplar: *eidos,* form, which is copied in accordance with the Idea."[41] Further distinctions follow. Thus: "an exemplar is internal (and formal), like the thought of the king's countenance or external, like the king's countenance."[42]

This is, once more, the "realistic" idea stemming from Seneca: there it was Virgil who was the model, and in this sense the idea the painter was copying. "It is natural, dianoetic: either efficacious, like a signet ring, or inefficacious, like the countenance of the king"[43]—presumably because the countenance doesn't produce anything. "It is eternal, like the exemplars of things in the divine mind, or perishable, like the type of the house in the mind of the architect. It is similar in species, like the shape of the king's countenance and the shape of the marble countenance. Or similar only in proportion, like the countenance of the king to the picture, the type of

---

38. Goclenius 1964, 428.
39. A further distinction reminiscent of Descartes's earlier work is that which follows between simple and complex ideas.
40. Crassot 1618, 104.
41. Crassot 1618, 105.
42. Crassot 1618, 105.
43. Crassot 1618, 105.

house to the house." In addition, "an exemplar is Platonic, or not: Platonic is either natural, like the horse itself, or mathematical, like the triangle itself, or artifactual, like the house itself, the shield itself, etc."[44]

The dianoetic exemplar is further specified: "The dianoetic exemplar is the form, which someone imitates. . . . The dianoetic exemplar is strictly an exemplar. But in the very strictest sense, it is that which is dianoetic and objective."[45] Here we have once more a possible source of Cartesian objective reality, this time, as with Goclenius's *conceptus,* specifically a distinction in type of concept. A dianoetic exemplar is "either an *objective* concept, like the king's countenance, or a *formal* concept, like the awareness of the king's countenance in the mind of the painter, or the dianoetic or intelligible *species* such as the *image* of the countenance of the king in the memory of the painter." As with Goclenius, the image sense enters here in the context of *species,* perhaps because the "species," the form without the matter, is what is taken up in perception and lingers as an image in the mind.[46]

The exemplar is also real or intentional, internal or external. Oddly enough, finally, "the thing thought seems to be able to be the exemplar of its future."[47] There is also a reference to God's making the thing itself through his thought. In short, the pair God and the artificer, eternal and perishable, creator and created, seems to be always invoked when exemplars/ideas are spoken of. In Descartes's case, however, as we have already noticed, the artist seems to be dropped.

The last example is Abra de Raconis. He is another of the authors whose name came up, along with Eustachius a Sancto Paulo, when Descartes wrote to Mersenne about the scholastic texts he might consult when he was considering drawing up a parallel set of principles of their philosophies and his.[48]

As usual, idea is identified with exemplar in the context of a discussion of causality: "exemplar or idea: for these two are the same."[49] And as usual the exemplar is what the artificer imitates "so that he can work according to the laws of his art." Again, also, a distinction is drawn between an external exemplar: "what is set before the eyes of the artist to be imitated," and an internal one: "what is formed by the artist in his mind."[50] Some people, de Raconis asserts, take the exemplar to be only internal, and in this he differs from them. On the existence of the exemplary cause we are given various

---

44. Crassot 1618, 105.
45. Crassot 1618, 105–6.
46. Compare with Hobbes, *Leviathan,* pt. 1, chap. 2: "imagination is nothing but decaying sense."
47. Crassot 1618, 106.
48. See Chapters 1 and 8.
49. Raconis 1651, 94.
50. Raconis 1651, 94.

authorities and reminded that, as St. Augustine has declared, the exemplar in the divine mind of things to be made is seen only by the blessed. As to exemplars in general, presumably in the finite (human) case, de Raconis provides a detailed discussion, clearly pertinent for the Cartesian enterprise, as to whether the exemplar consists in the objective or subjective (that is, formal) concept. De Raconis favors the former interpretation, which he supports with four propositions:

First proposition

The essential *ratio* of exemplar does not consist in the formal concept. . . .

This is demonstrated by common sense (*communi ratione*). The idea or exemplar is what the artificer contemplates in operating intentionally, as we plainly saw. But the artificer does not operate by contemplating his formal concept, for this inspection is a certain superfluous reflection of the mind, since without it the artificer can operate by contemplating the exemplar itself, or object of his formal concept. Therefore, etc.

Second proposition

The essential *ratio* of the idea or exemplar consists in the objective concept, or in the thing which is its object (*obiicitur*). . . .

. . . the exemplar is properly said to be what the artificer contemplates when he operates intentionally; and in fact the artificer operates by contemplating the objective concept of the thing to be made: e.g., the painter contemplating the objective concept of the image of Caesar, or the image itself, as it is projected on the mind (*obiicitur*), intends to paint another image similar to that.[51]

Here we meet "image" in both the senses anticipated in Goclenius's exposition: the image to be produced in addition to the image in the painter's mind. Clearly, it is the image itself as *obiicitur* in the mind that Descartes will draw on. The act of the maker (in Descartes's case, the thinker) is itself representative: it is *of* something reflected in the mind, and in this sense objective.

De Raconis reports and analyses several objections to this view and appends two further propositions in support of his own position:

Third proposition

It is not necessary that the exemplar be an objective concept distinct from the objective concept of the thing which is to be made by art; but it can be the concept of the thing itself to be made, known in advance by the artist.[52]

51. Raconis 1651, 95–96.
52. Raconis 1651, 97.

This proposition needs consideration particularly with respect to God's creation of things, since the ideas of them are contained in God's mind, yet they "are not like him, but imitate him only minimally."[53] He creates them through his idea of them, but they do not touch his essence. Another proposition follows:

Fourth proposition

> It is not necessary that the exemplar be the objective concept of the thing itself that is to be made, but it can be of something else in imitation of which the other thing was made.
>
> It is explained, for example: it can happen that the image of Caesar reflected in the mind (*menti obiecta*) might be the internal exemplar of another similar [thing] to be made, in so far as the painter looks at it assigning it to something similar. Therefore the exemplar can be the objective concept of another thing.[54]

Note the loosening here of the link between the painter's image and the reality. What is objectified in the mind has its own being, apart from Caesar himself, who is being copied. Analogously, the reality of a Cartesian idea, though it is *of* something, is itself a mental existent, not tied to an external reality. *Contra* Caterus (or Aquinas), it has its objective as well as its formal being. Finally, like our other writers, Abra de Raconis discusses the relation of exemplar causality to the four causes, and concludes that it is a kind of formal cause, but extrinsic. The exemplar/idea is not the "informing form" intrinsic to the thing but something outside it. Decidedly, however, it is not a fifth cause.

Now, this whole context is certainly not Cartesian. Although Descartes will link the objective reality of ideas to his version of the causal principle, the question for him is how ideas *are* caused, not how they operate as causes. He does not use the term "exemplar," which suggests the traditional Platonic context, where, as in the *Timaeus,* ideas are models or archetypes for creation. He does use the term "archetype" once, in Meditation III, when explaining that ideas, even if caused by other ideas, must ultimately by caused by something nonideal, "which is a copy of the archetype (*instar archetypi*) in which as much formal reality is contained as there is objective reality in the idea."[55] This does indeed echo the traditional

---

53. Raconis 1651, 97.
54. Raconis 1651, 98.
55. AT 7:42. CSM 2:29 translates the passage as "which will be like an archetype which contains formally <and in fact> all the reality <or perfection> which is present only objectively <or representatively> in the idea."

theme: if ideas are ultimately caused by things (Descartes's causal principle), those things copy the Ideas in the divine mind, which serve as models for them. But this is a very puzzling passage.[56] In general, however, Idea as model for creation, whether divine or human, is left aside in the Cartesian texts. What is retained is the notion of a mental act that is also (as we would say) intentional. And this Cartesian *leitmotif* does seem to be anticipated in de Raconis's account. The formal act of the mind, the particular thought ("as it were, image"), is *of* the object, though not necessarily of a "real" object "out there." The object of thought becomes, not necessarily the thing—Virgil, Caesar, the king's countenance—but what my mind seizes on in this thought (or "image") as its object. Thus in discussing ideas, de Raconis retains the traditional context of exemplary causality, while loosening the link between the formal, intellectual act and its object in a way reminiscent of the Cartesian position to come.

We have been looking at examples of the seventeenth-century use of the term "idea" with a view to finding background sources for Descartes's new, or reformed, conception, the idea that, for Arnauld and Nicole and their generation, would be self-evidently the unit of philosophical thought. What can we conclude from this brief survey?

First, clearly Descartes did draw on the current literary usage, in which an idea was a particular occurrent act of thinking, especially of imagination. And he had philosophical precedents for this as well, in writers like Eustachius, Goclenius, or Abra de Raconis. "Ideas are as it were images of things": they are mental acts that represent something: either my own thinking here and now (in the *cogito*, Meditation II) or God (in Meditations III and V) or the essence or existence of extended things (in Meditations V and VI respectively). Second, by calling on the ideas in *God's* mind as his source (rather than the minds of the painter or sculptor so frequently discussed) he set ideas as psychological entities free from their link to sensation: we noted earlier the frequent passages in which he insists on this distinction. Third, Descartes's insistence on the truth of ideas also has a precedent in the philosophical literature. The clear and distinct idea is the proper unit of thought, corresponding to the simple concept of Goclenius; and once we know that God is no deceiver, we can trust such luminous ideas to carry us to truth. Even though, in the terminology of the School, judgment is the formal locus of truth, for Descartes, in Meditations III and IV, judgment is the place where we may run the risk of error. It is in the as-

---

56. In the passage Descartes seems to be alluding to the tradition about ideas as exemplars—and even to ideas as images—only for the benefit of his audience, who would be well versed in the tradition. There is no reason to believe that these characteristics would be features of Descartes's ideas.

sertion (admittedly through judgment, especially through existential judgments) of our clear and distinct ideas, that, in one firm step after another, we proceed on the road to Cartesian science. Admittedly, Descartes has made an odd twist here in the relation of ideas to truth. In the usual discussion of exemplary (ideal) causality, it is things that must conform to the idea in the mind of the maker, divine or human. This accords with the account of the *Timaeus,* where Ideas serve as models for things as well as artifacts. The psychological idea, however, the particular mental act, can serve as vehicle for conformity *to* things rather than, conversely, as the standard to which things conform. Thus the adequacy of intellect to things can be transferred from risky judgments to much more reliable clear and distinct ideas. Ideas thus become, in Octave Hamlyn's words,[57] the *atoms of evidence* out of which argument, knowledge, the mind's road to reality, is to be built.

In taking this psychologizing path, however, Descartes seems (with the one exception of the archetype passage already mentioned) to have canceled out the major context of the traditional doctrine of ideas: the context of archetype or model, where the idea informs its imitations and gives them, or their "images," such reality as they have. Traditionally, ideas are identified with forms, species (eide): they have power through a certain agency; they are "efficacious," unlike particular things, which are relatively inert. It is Ideas that, primarily through God's thought (creation), inform things and make them the things they are. Analogously, though of course in a lesser degree, the artist's mind produces a copy of a reality, itself in turn informed by the divine patterns, the Ideas, in God's mind. But the artist is entirely ignored in Descartes's rendering, and even God as model follower occurs only in that one strange passage in the first proof of God. Nor does Descartes ever speak of "exemplars," which, though synonymous with "ideas," fail to carry the particular, psychological connotation implicit in the newer usage.

Why, we may wonder, did Descartes borrow the term "idea" but remove it almost altogether from its traditional context and import? Two reasons may be advanced, one epistemological and one metaphysical. Epistemologically, he was using the notion of God's ideas, which of course are not corporeal, to turn the new psychological meaning *away* from imagination to pure thought. His professed goal in the *Meditations* was to lead the mind away from the senses.[58] Cartesian ideas, aping God's pure cogitations, are entertained by minds qua minds, not by the embodied beings we now are. They are psychological units such as mathematicians employ in thinking

---

57. As quoted in Marion 1975, 135.
58. See Gueroult 1968.

through problems: thinking of thousand-sided figures as easily as of triangles and pentagons, which they could, but need not, imagine. Ideas as concepts, whether formal or objective or both in one, are the units by which Descartes (and his heirs) can free pure minding from its childish and scholastic bonds to sense (free it also from the foolish scholastic apparatus of judgments, syllogisms, and the like). Ideas, so isolated, thus perform an important epistemological function.

Metaphysically, moreover, when abstracted from their usual embedding in the Aristotelian tradition, they liberate the thinker from the whole rigmarole of form/matter ontology. We have, or enact, ideas of ourselves as thinking, of God as infinite power responsible for our very being through his creation and concurrence, and we have clear and distinct ideas also of matter as simply spread-outness. Thus, except in the case of our odd, if temporary, embodied being as mind united with body, we need no substantial forms, no quiddities, no real qualities: none of the rigmarole of scholastic definition and redefinition, argument pro and con, "more probable" or less probable views on these esoteric matters. Idea as a unit of thought, both formal and objective, frees the mind to move ahead in the construction of a science that is at least morally certain, and which the *Principles* will be meant to instantiate. It is indeed the case that Descartes made a new start in philosophy with his "idea," but it is also clear that he shaped this new conception by using readily available meanings of the term and at the same time purifying them of much of their habitual connotation.

[4]

The Cartesian Destiny of Form and Matter

with Marjorie Grene

What happens to the traditional form-matter pair in the Cartesian program? The answer at first sight appears simple. Scholastics saw everything hylomorphically —with the possible exception of the human rational soul between death and the last trump, and of course God. That is to say, they saw it as matter, itself mere potentiality, informed or actualized in such ways as to produce the nice, tidy variety of things, animate and inanimate, found around us. Within each specific form, matter is the principle of individuation, but always in subordination to a form. Descartes changed all that by making matter independent and replacing form, here and there only, by separate minds. So most of the erstwhile hylomorphic cosmos becomes just spread-out stuff, with minds dotted here and there as God decrees. This, it seems, is the Cartesian revolution.

When we look a little closer, however, we find a more complicated situation. Take the years 1637–41, between Descartes's first publication and the first edition of the *Meditations,* as roughly our *terminus a quo:* by then the "standard," if you will Thomistic, view of form and matter has been significantly modified, so that strict hylomorphism no longer prevails, and there are also some strident antischolastic voices to be heard, some, but not all, of these proposing corpuscularian themes. So we need first to look at the status of form and matter in the immediately pre-Cartesian literature. Second, we want to examine in Descartes's own writings such passages as shed light on his solution to the form-matter problem; and, third, we will consider the use of the concepts "form" and "matter" in the works of some adherents of Cartesianism: given that the gulf between the Scholastics and the *novatores* was not so great, we find here also a variety of compromises as well as some more radically antischolastic views.[1]

Before Descartes

Aristotle's doctrine, in *Physics,* book 1, is that there are three principles of natural things: matter, form, and privation. Though the principles are three, privation quickly drops out, being only incidental—"there is a sense in which the principles are two and a sense in which they are three," says Aristotle,[2] and Aquinas echoes: "There are two *per se* principles of the being and becoming of natural things, namely form and matter, and one *per accidens* principle, namely privation."[3] Questions usually arise about the relationship between the two *per se* principles of matter and form and their respective properties. Traditionally, matter and form are inseparable. All substances are informed matter. Form is associated with actuality and matter with potentiality: to be in actuality is to participate in a form and to have potentiality is to have a "power" of acting or undergoing something;[4] in this conception of substance, matter has the potential for receiving forms, whether substantial or accidental. Forms are kinds, or universals, and matter provides the individual substance with its particularity. Thus, matter is the principle of individuation, always subordinate to form, which makes it a this-such, a recognizable entity of such and such a kind. Substantial change, or mutation, that is, generation and corruption, is a change in the very nature of a thing, its acquisition or loss of a substantial form. Substantial forms are said to be indivisible, not capable of more or less, and not possessing contraries, and thus they cannot be acquired successively and piecemeal. Short of substantial change, motion, in contrast, occurs successively between contraries; motion must pass from one contrary as the term from which (*a quo*) to the other contrary as the term to which (*ad quem*). Since only forms in the categories quantity, quality, and *ubi,* or place, have

---

1. The thesis, then, is not that the seventeenth-century brand of Scholasticism directly influenced Descartes's formulation of his philosophy but that, at least, it prepared the way for the acceptance of Cartesianism (and for the eventual attempt at reconciliation). The thesis could be taken as a more general version of the one Vincent Carraud proposes with respect to Ockhamism and the reception of Cartesianism by Arnauld, in "Arnauld: From Ockhamism to Cartesianism," Ariew and Grene 1995, 110–28.

2. Aristotle, *Physics* I, 190b28–30. In fact, privation is needed only for substantial change, and so drops out of consideration for most of the *Physics*. It is commonly dropped by Cartesians, even those who still pay obeisance of a sort to form-matter explanation; see, for example, Rohault 1987.

3. Aquinas 1953, I, lectio 13.

4. Toletus 1589, III, chap. 1, text. 3.

contraries or positive opposite terms, Scholastics conceive of three kinds of motion: augmentation and diminution (in the category of quantity), alteration (in quality), and local motion (in place). A being moves, then, by virtue of the successive acquisition of qualitative or quantitative forms or of places. The association of matter with potentiality suggests that prime matter would be pure potentiality or nothing. In contrast, the association of form with actuality suggests that an ultimate form, or pure actuality, might subsist by itself.

The questions typically discussed in conjunction with these doctrines concern whether matter is a substance, whether potency is the essence of matter, whether matter is not capable of being generated or corrupted, whether matter is disposed to receiving the form, whether matter or form is the cause of corruption,[5] whether some forms preexist in matter, in what way form arises from matter, whether forms can be outside matter,[6] and, ultimately, whether there can be any prime matter separate from forms.[7] There is much agreement and disagreement in the answers given to these questions. One can point to almost universal agreement among late Scholastics concerning the negative answer to the question of whether forms are generated from matter. Although late Scholastics usually repeat the phrase "form results from the potentiality of matter, that is, from the natural aptitude of matter to receive various forms in succession," they do not understand it as indicating that form receives its nature from matter.[8] Similarly,

---

5. For example, Toletus 1589, quaest 13, "An materia sit substantia"; quaest. 14, "An potentia sit de essentia materia"; quaest. 16, "An materia sit ingenerabilis et incorruptibilis"; quaest. 17, "An materia appetat formam"; quaest. 20, "An materia sit cause corruptionis, an forma." Eustachius a Sancto Paulo 1629, pars 3, bk. 1, disp. 2, quaest. 2, "Quid sit materia, et quare sit admittenda"; quaest. 3, "Quaenam et qualia sit potentia materiae"; quaest. 4, "Quaenam sint praecipuae proprietates materiae." Raconis 1651, pars 3, Physica, Tractatus de Principiis, disp. 2, art. 3, memb. 5, "Utrum materia sit pura potentia Metaphysica"; memb. 6, "Utrum materia sit pure potentia physica." Many of these issues are discussed in detail in Des Chene 1995b.

6. For example, Toletus 1589, quaest. 19, "An aliquid formae praefuerit in materia." Eustachius a Sancto Paulo 1629, quaest. 6, "Quomodo ex materia nascantur formae"; quaest. 7, "Quomodo ab agente producantur formae"; quaest. 9, "An formae extra materiam esse possint." Raconis 1651, memb. 2, "Utrum potentia materiae ad eius essemtiam spectet"; memb. 3, "Ad quas formas se extendat materiae potentia"; memb. 4, "Utrum illa materiae potentia prius respiciat formas substantiales, quam accidentales."

7. For example, Dupleix 1990, chap. 5, "Resolution des argumens qui concluent qu'il n'y peut avoir de matiere premiere separée des formes."

8. See, for example, Dupleix 1990, bk. 2, chap. 2; Eustachius 1629, III.1.2, quaest. 6–7. See also Sennert 1659, bk. 1, chap. 3. Sennert goes a step further than the usual textbook writer. He accepts the standard account, that neither matter nor form are generated, only the compound of matter and form, but when he comes to the scholastic phrase that forms are drawn out of the aptitude or potentiality of matter, he says that he hears the sound of the words, but that his mind hears nothing. Ultimately he accepts Toletus's rejection of the opinion that there was something of form in matter before its introduction therein.

seventeenth-century Scholastics agree that at least one form can subsist without matter, namely rational soul.[9]

But can matter exist without form? As we shall see, this is a crucial question for Cartesianism, since a positive resolution of this esoteric question might lead one toward a dualistic, as opposed to a hylomorphic, conception of substance. At times, there is sharp disagreement on the answer. According to Aquinas, prime matter is pure potency, or has only potential being;[10] thus prime matter was not brought into being without form, and matter cannot subsist without form.[11] Scotus objects; he holds matter to be a positive entity really different from the reality of form, which can subsist in its own right distinct from form.[12] The motivation for the position seems to have been his wish to preserve God's absolute omnipotence as far as he could. Scotus claims that God can create matter without any form, whether accidental or substantial: "Every absolute thing that God produces among creatures by the intermediary of a second cause, he can create without this second cause, which is not part of the effect. Now, the form that confers existence on matter is a second cause and is not part of the essence of matter insofar as it is matter. Hence God can create the matter without the form."[13]

Franciscus Toletus knows both positions. In question 13 of his *Commentary on the Physics,* book 1, he discusses whether prime matter is a substance; he details both Scotus's affirmative reply to the question and Aquinas's negative answer—that prime matter is pure potency—in order to side with Aquinas. Toletus's own doctrine is that prime matter is imperfect in itself.[14] Toletus then discusses whether matter can exist without form. He refers to Aquinas's thinking that that would be impossible, since it would imply a contradiction,[15] and to Scotus's doctrine that it can be done by supernatural means (but without giving references to Scotus). He concludes that he sides with Aquinas, that there cannot be any matter in act without a form. Against Scotus he again argues that matter is imperfect in itself.[16] Others, such as Théophraste Bouju, also followed the Thomist line about the reality of prime matter.[17]

---

9. Support for this position comes from *De anima* III, chap. 5, concerning the active intellect.
10. Aquinas 1964–76, I, quaest. 7, art. 2.
11. Aquinas 1964–76, I, quaest. 66, art. 1.
12. Scotus 1968, *Opus Oxoniense* II, dist. 12, quaest. 1. William of Ockham also holds a similar doctrine.
13. Scotus 1968, *Opus Oxoniense* II, dist. 12, quaest. 2.
14. Toletus 1589, quaest. 13, "An materia sit substantia," fol. 34v.
15. Toletus 1589, fol. 35r: "Sanctus Thom. I. p. q. 66. ar. 1. & quodlibeto 3. arg. 1."
16. Toletus 1589, fol. 35r.
17. See Bouju 1614, 1:315–16 (chap. 6, "Que la première matière est pure puissance passive, et comment"); 319–20 (chap. 11, "Comment la première matière est moyenne entre l'estant et le non estant"); 322 (chap. 15, "Comment la forme donne l'estre au composé");

On the other hand, Eustachius a Sancto Paulo's doctrine differs from Toletus's. As usual, there is no difficulty about forms existing without matter in the case of the rational soul.[18] The main problem arises with respect to matter's existing without any form. Eustachius supports a variant of Scotus's doctrine, though without citing sources, thus without naming Scotus or mentioning Aquinas's doctrine that prime matter is pure potency: "Although matter cannot be produced nor annihilated by any natural agent, God can create or annihilate it. . . . God can strip naked all forms, substantial and accidental, from matter, or create it naked, without form, ex nihilo, and allow it to subsist by its own power in such a state."[19] This looks, *a tergo*, like a lead-in to Cartesian dualism; it is certainly a weakening of the traditional close linking of matter to form. Moreover, de Raconis agrees; he quotes both Thomas Aquinas and Duns Scotus, says that matter is an incomplete substance, but maintains that God can create matter without substantial form.[20]

Scipion Dupleix puts into relief the disagreement between Thomists and Scotists:

> Thus matter deserves the name of substance because it subsists by itself and is not in any subject. This reply is based on the Philosopher's doctrine, but it does not satisfy everyone, particularly Saint Thomas Aquinas and his followers, who hold that such matter is not in nature, and cannot be in it, and even that this is so repugnant to nature that God himself cannot make it subsist thus stripped of all form. But this opinion is too bold, too mistaken, and as such it has been rejected by Scotus the Subtle [Doctor] and by several others who convicted Saint Thomas by his own words.[21]

It is interesting to note that Dupleix argues against Aquinas's doctrine of prime matter by analogy to what is required in the sacrament of the Eucharist.[22] For his purpose, Dupleix only needs to argue against the

---

326–27 (chap. 23, "Que la nature et forme naturelle ne se trouvent jamais séparées naturellement l'une de l'autre"); 329–30 (chap. 26, "Réfutation d'une prétendue puissance objective en la première matière, et de l'acte objectif qui luy respond"); 330–31 (chap. 27, "Rejection de l'acte entitatif ou objectif, que quelques uns ont estimé estre en la première matière"); and 331 (chap. 28, "Réfutation de l'opinion que la première matière peut estre naturellement sans la forme").

18. Eustachius a Sancto Paulo 1629, pars 3: Physica, bk. 1, disp. 2, quaest. 9: "An formae extra materiam esse possint," pp. 22–23.

19. Eustachius 1629, quaest. 4: "Quaenam sint praecipuae proprietates materiae," pp. 16–17.

20. Raconis 1651, Tractatus de Principiis, disp. 2, memb. 4, "Utrum materia sit pura potentia metaphysica," pp. 35–39.

21. Dupleix 1990, 131.

22. "[Saint Thomas] accorde bien que Dieu peut faire que l'accident subsiste en la nature hors de son sujet: comme mesme tous les vrays Chretiens croyent que tous les accidens du

Thomists that God is able to create matter without form, but he goes further. He asserts that matter subsisting without form "is not repugnant to nature and still less to divine power, which is infinite and above everything in nature." He adds that "even though matter is not found separated from forms, it is nevertheless something distinct and separate from form in essence, and it even precedes form when one considers the generation of natural things."[23] The fine line Dupleix wishes to draw is plainly exhibited when he considers the creation of matter and form. He states that there is never any matter without form in nature but that we can conceive matter without form without in any way upsetting the natural order:

> in the same way that we ordinarily consider the virtues, vices, colors, dimensions and other accidents outside their subject, even though they are never separated from it, [we can consider] substances without having any regard to their accidents, which can be elsewhere than in them. That is why the ancient Pagans did not recognize that God created this matter as well as the forms at the beginning of the world, and thinking instead that it was something separate from forms, they imagined a chaos, a confused and inform mass corresponding to this prime matter, from which they made all things arise.[24]

Dupleix then cites verses of Ovid about chaos and lack of order at the creation and even suggests that Moses himself followed this "natural order," representing prime matter at the beginning "as the principle of all created things" by the words darkness (*tenebre*), waters (*eaux*), abyss (*abysme*), and void (*vuide*).[25] But Dupleix's doctrine is clear: matter can exist without form naturally and by supernatural action; we can conceive it thus; but it simply does not so exist, given that it was created simultaneously with form. Still, it could!

If we were to jump forward to the second half of the seventeenth century, we would find that even then not all philosophers accepted the Cartesian position on these questions. Schoolmen were still disputing the same issues as Toletus versus Eustachius or Dupleix. And not all textbook writers went as far as Dupleix. Some just accepted the reality of matter as a miracle—for example, René de Ceriziers argued that there can be no form without mat-

---

pain sont au S. Sacrament de l'Eucharistie sans le pain: et les accidens du vin sans le vin: bien qu'il semble y avoir beaucoup plus de repugnance en cecy qu'a faire subsister la matiere sans forme: d'autant que la matiere n'a pas besoin d'aucun sujet ny de suppost, estant elle mesme le sujet et le suppost de toutes autres choses naturelles: et que l'accident ne peut naturellement subsister sans sujet." Dupleix 1990, 131–32.

23. Dupleix 1990, 132.
24. Dupleix 1990, 130.
25. Dupleix 1990, 131. One can find the same creation story of simultaneous matter and form in Sennert 1659, book 1, chap. 3, and then again in book 9, chap. 3.

ter and no matter without form by natural means, but added, "however, one must not deny that God can conserve matter without any form."[26] This compromise solution seems to have been unstable, so in 1665 the Jesuit Pierre Gaultruche argued against the Thomists (*contra Thomistas*) about prime matter.[27] Not everyone gave up the Thomist doctrine of matter, however. Although Scotists such as Claude Frassen seem to have had the best of the argument, and Thomists and Jesuits such as Pierre Barbay and Jean Vincent needed to opt for a middle ground, some Thomists resolutely maintained their position.[28] For example, the Dominican Antoine Goudin wrote: "it can be asked whether God by means of his omnipotence could create matter without its having a form. Scotus asserts it, as do some authors outside of Saint Thomas' school; Saint Thomas and all the Thomists deny it. . . . It seems that matter cannot exist without form even by means of God's absolute power. That is what Saint Thomas states (III quodlib. art 1). God himself cannot make it that something exist and not exist. He cannot make something that implies a contradiction and, consequently, he cannot make matter be without form."[29]

To return to the period we were considering: as the dominant scholastic position became somewhat more dualistic than hylomorphic, with matter being endowed with being, another trend—perhaps in the opposite direction—was the shifting of one of the principal functions of matter to form. The principle of individuation became form instead of matter, with consequent changes in what is meant by form. We can grasp the radical change when we read Dupleix's exposition of form in his *Physique*. The question Dupleix wishes to answer is why there is not a prime form common to all matter, as there is prime matter that is common to all forms. His answer is that "form is not only that which gives being to things, but also that which diversifies and distinguishes them from one another. Thus, nature, which is pleased with diversity and variety, cannot allow there to be a form common to all matter, as there is a matter common to all forms; if there were only a single form, as there is a single matter, all things would not only be similar, but also uniform and even unitary."[30]

Dupleix's discussion of the problem in his *Métaphysique* is exemplary. He points out that there are three main opinions about the principle of individuation: the Thomist, the Scotist, and that of another group he does not identify. Thomists think the principle of individuation is signate matter,

---

26. De Cerizier 1643, chap. 3, pp. 51–52.
27. Gaultruche 1665, vol. 2, Physica Universalis, p. 27.
28. Frassen 1686, 36–41; Barbay 1676, *Physica*, pp. 64–72; Vincent 1660, 2:74–77.
29. Goudin 1864, vol. 2, art 4, p. 131.
30. Dupleix, 1990, 135.

which they understand in a variety of ways: matter limited to a certain quantity by its dimensions; a composite of matter and quantity; or the power and aptitude of matter to a certain limited quantity.[31] Dupleix points out that the Thomists have a difficulty with the principle of individuation for noncorporeal substances such as angels, whose principle is based on their lack of matter (every angel or intelligence must be considered as both universal and individual).[32] The anonymous group consists of those who base the principle of individuation on the "multitude of accidents," given that this multitude "is never found together in any other subject."[33] The Scotists reject the other two opinions and maintain that the principle of individuation must be based on an ultimate specific difference (*haecceitas*) for each individual. Dupleix allows that the Thomists (especially those holding the first interpretation) have the authority of Aristotle behind them,[34] but he does not think that they are right. He agrees that quantity can be a mark of individuation for corporeal substances, but he does not think that it reveals "the proximate and true formal cause of the individuality and unity of the essence of singular things,"[35] since quantity is always an accident and accidents do not operate at the level of essences. He repeats this argument against the second, anonymous opinion; and he dismisses the other groups of Thomists with roughly similar arguments: that specific difference is universal and cannot be both principle of individuation and of universality.[36] Dupleix's preferred position is the Scotist opinion that "in order to establish the individual essence of Socrates, Alexander, Scipion, and other singular persons, we must necessarily add for each one of them an individual and singular essential difference which is so proper and so peculiar to each of them for themselves, that it makes each of them differ essentially from all the others."[37]

There are similar doctrines in the Metaphysica of Eustachius a Sancto Paulo and Abra de Raconis.[38] Another doctrine in the same general direc-

---

31. Dupleix 1992, 231–32.
32. Dupleix 1992, 232.
33. Dupleix 1992, 232. Cf. Leibniz on the principle of individuation.
34. "Pour le regard du premier chef consideré en gros et en general sur la matiere, il semble à la verité estre fondé sur la doctrine du Philosophe, lequel establit quelquefois l'estre des substances corporelles en la matiere: comme quand il dit au 5. livre de sa Metaphysique que *ces choses-là sont en nombre, desquelles la matiere est une:* Et au livre 7. *que la chose singuliere de la derniere matiere c'est desja Socrates,* c'est à dire, un individu" (Dupleix 1992, 233).
35. Dupleix 1992, 233.
36. Dupleix 1992, 234.
37. Dupleix 1992, 235.
38. Raconis 1651, pars 4, Metaphysica, tract. 4, sec. 2, 4, brevis appendix, "De unitate singulari et numerica, seu principio individuationis," pp. 76–78; Eustachius a Sancto Paulo 1629, pars 4, Metaphysica, tractatus de proprietatibus entis, disp. 2, De simplicibus proprietatibus Entis, quaest. 4, "Quodnam sit principium unitatis numerica, seu individuationis," pp. 38–39.

tion is that of Franco Burgersdijk in his *Institutionum metaphysicarum* (originally published in Leyden, 1640).[39] Burgersdijk rejects both Scotus's and Thomas's opinions (with Thomas's as the worse one).[40] His own doctrine is that composite substances have both material and formal principles of individuation. With humans, the individuality lies in the rational soul, which is an immaterial form. And, of course, humans can also be differentiated by their accidents.[41]

The Scotist position seems to be the majority position in the seventeenth century. It entails that form is the principle of individuation. This appreciably alters what one means by form; forms are no longer necessarily specific. Thus form is on its way to becoming just the way a particular part of matter is differentiated: ultimately, structure or shape, rather than the organizing principle that makes the thing the kind of thing it is.[42]

So much for the changes in scholastic thought: generally speaking, the strengthening of matter and the weakening of form. At the same time, there were also significant antischolastic voices, the *novatores*, who included a variety of physicians and alchemists. Among the former, one can count Sebastian Basso (and Daniel Sennert) and, among the latter, Etienne de Clave. In 1630, Descartes remarked to Beeckman of such writers: "Plato says one thing, Aristotle another, Epicurus another, Telesio, Campanella, Bruno, Basso, Vanini, and all the innovators (*novatores*) all say something different. Of all these people, I ask you, who is it who has anything to teach me, or indeed anyone who loves wisdom?"[43] Yet, however he may have sneered at them, the innovators were in fact preparing the way for Descartes's "revolution."

Of the *novatores* Descartes listed, Basso, it is established, is the one whose work he knew.[44] In humanist fashion, Basso wants to recover the philosophy of the ancients previous to Aristotle, and in particular atomism. For him, the ultimate constituents of bodies are the minimal particles of matter or atoms. Each atom is homogeneous, a simple body possessing a particular nature that persists in mixtures;[45] when atoms enter into composition, they make up natural minima having their own proper natures.[46] According to

---

39. Burgersdijk 1657, I, chap. 12, "De Unitate Numerica et formali, deque principio individuationis," pp. 66–75.
40. Burgersdijk 1657, 71–72.
41. Burgersdijk 1657, 74–75.
42. For a similar movement in the conception of "idea," from ideas as Forms to ideas as particular images, see Chapter 3.
43. To Beeckman, 17 October 1630, AT 1:158.
44. See AT 1:25, 665. See also Chapter 6.
45. Basso 1649, 27; see also the resumé, p. 67.
46. Basso 1649, 23.

Basso, there are four kinds of elementary atoms (other than the ether), coinciding with the four traditional elements. But Basso contests the scholastic doctrine that the four elements can assume new substantial forms and thus can be generated from one another.[47] Indeed, for Basso, all change—generation and corruption, alteration in quality, and augmentation and diminution in quantity—is explicable at the level of the ultimate constituents of matter. Generation and augmentation in quantity are the gathering together of atoms or clusters of atoms. Corruption and diminution in quantity are the dispersing of atoms that were previously united. Alterations in quality result from atoms of one kind being substituted for atoms of another.[48] Thus, for Basso, completely new generation is an illusion; what happens instead is the continuous reorganization of atoms.[49] Although Descartes was of course no atomist, the transformation of generation into mere change of place or shape would also hold for his world of *res extensae*.

Further, Basso's concept of causation is mechanistic (or, more accurately, proto-mechanistic), not, however, in Democritean fashion, through motion in a void, which he rejects, but through a fifth element, the ether, far more tenuous than the elementary atoms and permeating every kind of object insofar as it fills the gaps between the atoms of the four elementary kinds.[50] It is the cause of the motion of atoms and, in this way, the cause of the mutations of bodies.[51] Yet although the ether is the cause of motion, it is totally inert in itself, and in constant need of being kept in motion by a higher cause. Basso refers to a kind of Platonizing universal form; ultimately, God is the higher cause on which the ether depends, not only for its motion but also for its directing of the motion of the elementary atoms. Although Descartes questions Basso's use of the ether (in the particular context of rarefaction and condensation),[52] the physician's reduction of all kinds of change to the local motion of atoms—with God coming, not from, but into the machine—does seem congenial to the Cartesian view.[53]

---

47. Basso 1649, 118 et seq.
48. Basso 1649, 72.
49. Basso 1649, 9–10.
50. Basso 1649, 304, 306.
51. Basso 1649, 308–9, 387–88.
52. See Descartes to Mersenne, 8 October 1629, AT 1:25.
53. We should note that Basso's atomism and anti-Aristotelianism make him a heretic for some early seventeenth-century thinkers (and not only for the most conservative ones). Discussing "Gorlee, Charpentier, Basso, Hill, Campanella, Brun, Vanini, et quelques autres," Mersenne complains of their "impertinence" and denounces atomism, that is, the doctrine "qu'il y a des atomes dedans les corps, qui ont quantité et figure"; according to him, "en bout du conte ils sont tous Heretiques, c'est pourquoy il ne faut pas s'estonner s'ils s'accordent comme larrons en foire." Mersenne 1624, 237–38. See Chapter 6.

Etienne de Clave is another early anti-Aristotelian. Although Descartes does not refer to him in his published correspondence, there is no doubt that he knew of his opinions, since they became notorious—the subject of denunciations by the circle around Descartes—in the middle 1620s (a time when Descartes was in Paris and in contact with Mersenne and others).

De Clave, Jean Bitault, and Antoine Villon scheduled a disputation for August 24 and 25, 1624 by posting a broadsheet containing fourteen anti-Aristotelian theses on the streets of Paris. The disputation did not take place. The president of the Parlement saw copies of the theses and prohibited the disputants from sustaining them on pain of death. The Parlement then sent the theses to the Faculty of Theology of the University of Paris (the Sorbonne) to be examined. A few days later, the Sorbonne replied with a censure of some of the theses and, through an *arrêt* of 4 September 1624, the Parlement ordered Villon, de Clave, and Bitaud to leave Paris, never to teach again within their jurisdiction, on pain of corporal punishment.[54] Among the prohibited theses were propositions concerning matter and form, one in particular denying all substantial forms, except for rational soul, along with prime matter; its official condemnation is that "this proposition is overly bold, erroneous, and close to heresy."[55]

There are many extant reports about the event of 1624, including some by Descartes's correspondents Mersenne and Jean-Baptiste Morin, as well as by others such as J.-C. Frey, professor of philosophy at Paris. These reports have little favorable to say about the theses of de Clave, Bitaud, and Villon. Mersenne defends Aristotle against their attacks and dismisses them as charlatans.[56] He goes through all fourteen posted theses,[57] expressing

---

54. See Garber 1988, 471–86; 1994.
55. "Formae item omnes substantiales (excepta rationali) non minus absurde defenduntur ab Aristotelicis quam materia, cum per eas intelligant substantias quasdam incompletas, unum per se cum materia substantiale compositum constituentes; materia enim e naturali composito sublata, et formas saltem materiali tolli necesse est. CENSURA: Haec propositio est temereria, erronea, et haeresi proxima" (Launoy 1656, 310–21). This prohibition was renewed in 1671 and became the basis for condemnations of Cartesianism; see Chapter 9.
56. Mersenne 1625, 100–101.
57. The anti-Aristotelianism seems to derive from an alchemical and atomistic bent. Here is Mersenne's description of theses 1–8 and 12–14:

> Il me semble qu'elles s'opposoient particulierement à la doctrine d'Aristote, et que les deus premieres nioient la matiere, et la forme: la troisieme se mocquoit de la privation: la 4. et la 5. vouloit que chaque mixte fut composé de cinq corps simples, sçavoir est de terre, d'eau, de sel, de souphre, ou d'huile, et de mercure, ou d'un esprit acide, lesquels la 6. these asseuroit être d'une même espece en tous les individus, de maniere que la diversité de tous les genres, des especes, et des individus ne provenoit d'ailleurs que du divers mélange de ces principes, qui étoit encore cause selon la 8. these, de toutes les actions, et de tous le mouvemens qui se voyent dans tous les individus sensibles en ce monde ... La 12. ôtoit les qualitez virtuelles, et rapporatoit tous les effects qu'on attribuë ordinairement a ces qualitez, au 5 substances qu'ils se mettent en chaque corps:

general disapproval. His main argument is that "if there is no form and no matter, then man has neither body nor soul, something contrary to the belief of the Catholic faith; if there are no other genera and no other species, except for the various mixture of the five substances established by them, man is of the same species as stones, plants, and animals, which is most false."[58] Morin seems to take as basic and beyond question the Aristotelian view that "matter . . . and form united are the essence of body as such." He then argues that without matter and form, there can be no bodies—there cannot even be a human body for a soul to inform, since the body without its own form is nothing.[59] In a similar vein, working on the lack of parallelism between humans and other animals (humans having rational souls, but animals lacking any substantial form), Frey asks rhetorically: If a donkey is a donkey without the substantial form for being a donkey, then why would a human not be human without the substantial form for humanity? And if a human is formally a human by its substantial form, why would a donkey not be a donkey by its own substantial form?[60]

De Clave's banishment did not prohibit him from subsequently publishing a number of alchemical treatises consistent with his earlier anti-Aristotelian views. As becomes evident in the later work, however, the denial of substantial forms does not entail the denial of form-matter talk. De Clave continued to reject the Aristotelian doctrine of four elements (he denied the element fire)[61] and the doctrine that the elements were the product of permutations of the opposing qualities hot, cold, dry, and moist.[62] He held that a primary element can only be derived from a single form and that form could not itself be derived from matter.[63] He denied substantial forms for things such as donkeys, preferring to think of them as composites of primary elements, themselves considered as form-matter.[64] For all the noise de Clave, Bitaud, and Villon made, their rejection of form was only partial.

---

le 13. se mocquoit de toutes les transmutations qui se font entre les élements, et soutenoit que la terre ne peut jamais être changée en eau, ni l'eau en terre, ny l'un des 3. principes en un autre: d'où ils concluent en la 14. qu'Aristote avoit tort de s'être mocqué de ces deus maximes des anciens sçavoir est que toutes choses étoient en toutes choses, et que toutes choses etoient composées d'atomes." (Mersenne 1625, 79–80)

58. Mersenne 1625, 81–82.
59. Garber 1994.
60. Frey 1628, chap. 27.
61. De Clave 1641, chap. 1, "Du nombre des elemens Peripatetiques."
62. De Clave 1641, chap. 13, "Des qualitez elementaires."
63. "Mais dautant que les proprietez sont plus intrinseques, comme estant dependentes immediatement de la forme, il est raisonnable de commencer par elles . . . il n'y a aucune apparence de dire qu'elles procedent de leur matiere, puis qu'elle n'a aucune qualité, estant informe voire un estre incomplet, qui ne reçoit aucune resistence, moins encore de perfection, avant que d'estre informée, c'est à dire, avant de la venuë de la forme qui lui donne estre et perfection" (Clave 1641, 117).
64. See Emerton 1984, 60–61.

Other writers also helped to shatter the traditional reliance on form as an explanatory concept. Two who influenced Adriaan Heereboord, the first Cartesian in Leyden, for example, were the brothers de Boot (or Boate), who published a *Philosophia Naturalis Reformata* in Dublin in 1641, and William Pemble, a Fellow of Magdalen College, Oxford, whose *De Formarum Origine* first appeared in 1628.[65] Both these writers raise troubling questions about the meaning of "form." Forms turn out to be neither corporeal substances, nor immaterial substances, nor "temperaments" of substances, and so on. Are they perhaps just accidents of bodies? All this always after having made an exception of the rational soul, of course. We are clearly moving in a decidedly Cartesian direction.

Descartes

If, on various fronts, the way had been prepared for his new program, that is not to deny that Descartes himself made, and encouraged in others, a radical break with the hylomorphic tradition. In his mature work, he unequivocally elevates matter to the rank of substance and emphatically eliminates the various kinds of soul that used to mediate between mere matter and separable mind. There is finite extended substance (which can stretch indefinitely, but is not infinite) and finite thinking substance and God, or infinite spiritual substance, and that is that. Form-matter thinking, even the problem of form and its origin, which worried people like Pemble or the Boots, seems to have faded from view. In the *Meditations*, his major argument on "first philosophy," Descartes's simply does without a form-matter perspective altogether. There is, of course, the one case of our minds' informing our bodies during our sojourn in this vale of tears: this is even, temporarily, a *substantial* form, like those the Scholastics had foolishly attached to other bodies. For them, mind had been the only entity separable from matter; for Descartes it is the only substance that does, in our case, function as the form of another substance, the human body, or, if you like, as the substance informing another substance to constitute a complex substance.[66] Over against his insistence on this one, exceptional substantial form, Descartes does indeed attack the notion of substantial form directly on a number of occasions, while in the *Meteors* he remarked that although he had nothing against the concept, he just didn't need it.[67]

---

65. Boot and Boot 1641; Pemble 1650.
66. See AT 7:434; CSM 2:293. See also Grene 1991, 38–39.
67. "Puis, sçachés aussy que, pour ne point rompre la paix avec les Philosophes, je ne veux rien du tout nier de ce qu'ils imaginent dans les cors de plus que je n'ay dit, comme leurs *formes substantielles*, leurs *qualités reelles*, et choses semblables, mais il me semble que mes

In an admonitory letter to Regius, whose statement that man was a being "per accidens" was instrumental in fueling the notorious quarrel between Descartes and the Utrecht theologians, Descartes called Regius's attention to that remark in the *Meteors*. Adjuring him not to use harsh words, he continued:

> I should like it best if you never put forward any new opinions, but retained all the old ones in name, and merely brought forward new arguments. . . . For instance, why did you need to reject openly substantial forms and real qualities? Do you not remember that on p. 164 of my *Meteors*, I said quite expressly that I did not at all reject or deny them, but simply found them unnecessary in setting out my explanations? If you had taken this course, everybody in your audience would have rejected them as soon as they saw they were useless, and in the mean time you would not have become so unpopular with your colleagues.[68]

This piece of advice is revealing for Descartes's own use of traditional concepts, and of "form" in particular. Matter, of course, has been officially liberated (except in the case of humans) from its hylomorphic captivity, and reigns as *res extensa* throughout the natural world. Form, on the other hand, seems at first sight to have slipped quietly away. There are of course those passages, chiefly in letters or in the *Replies*, in which, contrary to his own shrewd advice, he explicitly takes issue with the notion of substantial form.[69] However, if we put those to one side, and look at Descartes's use of "form" as such, we find some revealing instances of the technique he had recommended to Regius: turn the old ways to your own use.[70] Let us go quickly through some examples.

To begin with, in the early *Olympica*, Descartes remarks that "every corporeal form acts through harmony."[71] This is not the Descartes we know; it shows us, however, the young Descartes as deeply the product of a scholastic education. This is the kind of thinking he needed to recover from—and he did so brilliantly.

The unpublished *Le Monde* shows the way he is going: "Others may, if they wish, imagine the form of fire, the quality of heat, and the process of burning to be completely different things in the wood. For my part, I am afraid of mistakenly supposing there is anything more in the wood than

---

raisons devront estre d'autant plus approuvées, que je les feray dependre de moins de choses" (AT 6:239).

68. AT 3:491–92; CSM 3:205 (with revision).
69. For such passages, see Grene 1991, esp. 17–19.
70. True, in this case we are speaking of a term, "form," not of an opinion, but it was a term that was centrally used in conveying the old opinions and would have been comfortably familiar to Descartes's readers.
71. Descartes, *Olympica*, AT 10:218; CSM 1:5.

what I see must necessarily be in it, and so I am content to limit my conception to the motion of its parts."[72] And he goes on, not to banish the word "form," but to use it in his own way: to mean "shape and size," with local motion as their cause: "the forms of inanimate bodies . . . can be explained without the need to suppose anything in their matter other than the motion, size, shape, and arrangement of its parts."[73] So he continues, throughout the text, to use "form" to mean "shape and size," in particular, to refer to the "forms" of his three elements, but also more generally, to the character of mixed bodies, or even of "a quite perfect world."[74]

At the same time, there is one passage in which he seems to be using the term "form" in a more traditional way, in connection with his introduction of extension as the basic characteristic of matter. The philosophers, he says, "should . . . not find it strange if I suppose that the quantity of the matter I have described does not differ from its substance any more than number differs from the things numbered. Nor should they find it strange if I conceive its extension, or the property it has of occupying space, not as an accident, but as its true form and essence."[75] Here we see Descartes struggling with the terminology needed to carry his new conception of nature. Later, in the *Principles*, extension will be labeled the "principal attribute" of matter, and, as we shall see, "form" will unequivocally denote shape and size, the products of local motion. To someone used to scholastic ways of thinking, the notion of extension, the plain taking up of space, as the "true form and essence" of anything is curiously unsettling.

Where else does Descartes use the concept form? There is a surprising passage in part 1 of the *Discourse*, where he is insisting that "reason and sense" are equally present in all men. "Here," he says, "I follow the common opinion of the philosophers, who say there are differences of degree only between the accidents, and not between the forms (or natures) of individuals of the same species."[76] This is following his advice to Regius with a vengeance, using the terminology of his teachers to introduce an argument that will lead in a direction diametrically opposed to theirs.

So far as we have discovered, however, a positive use of "form" occurs again only in the *Principles*, when Descartes has shaken his own ontological terminology into shape. Here, again, it is by no means a central concept,

---

72. Descartes, *Le Monde*, AT 10:7; CSM 1:83. In another passage he also explicitly rejects the notion of a "motus ad formam," along with other scholastic conceptions of other than local motions; AT 10:39, CSM 1:94.

73. Descartes, *Le Monde*, AT 10:25; CSM 1:89.

74. Descartes, *Le Monde*, AT 10:26, 33, 34, 39, 48, 51–52; CSM 1:89, 91, 94. Descartes 1979, 79, 83–85.

75. Descartes, *Le Monde*, AT 10:35; CSM 1:92. This passage deserves to be quoted, and studied, more fully than we can afford in the present context.

76. Descartes, *Discours de la methode*, AT 6:2; CSM 1:111.

but it occurs in two passages, in both of which it has the meaning of figure or size that it had already acquired, to a large extent at least, in *Le Monde*.[77] *Principles* II, article 23, tells us: "*All the variety in matter, all the diversity of its forms, depends on motion.*" And the explanatory paragraph shows us in a nutshell how satisfactorily Descartes has come to terms with scholastic form-matter talk:

> The matter existing in the entire universe is thus one and the same, and it is always recognized as matter simply in virtue of its being extended. All the properties which we clearly perceive in it are reducible to its divisibility and consequent mobility in respect of its parts, and its resulting capacity to be affected in all the ways in which we perceive as being derivable from the movement of the parts. If the division into parts occurs simply in our thought, there is no resulting change; any variation in matter or diversity in its many forms depends on motion. This seems to have been widely recognized by the philosophers, since they have stated that nature is the principle of motion and rest. And what they meant by "nature" in this context is what causes all corporeal things to take on the characteristics of which we are aware in experience.[78]

A brief reference to "forms" in the text of *Principles* II, article 47, though incidental to Descartes's argument there, is perhaps equally revealing. By the operation of the laws of nature, Descartes tells us, "matter must successively assume all the forms of which it is capable."[79] As Vincent Carraud has pointed out, Descartes is here echoing a statement by Saint Thomas, but with a very different intent, since here, again "form" is simply shape and size, and it is through Descartes's (divinely decreed) laws of (purely local) motion that the "successive assumption" of all forms possible for it will occur.[80] The same words, with a wholly different intent. Did they convince his readers, as he hoped they would? The reactions of those he did not con-

---

77. We are ignoring adjectival or adverbial uses of "formal," "material, "formally, "materially," which of course occur frequently even in the *Meditations*. It would be difficult to speak to scholastic readers, as Descartes was doing, without using such terms; they do not materially (or perhaps we should say "formally"!) affect Descartes's argument.

78. AT 8A:52–53, CSM 1:232–33.

79. AT 8A:103, CSM 1:258. Descartes had said the same thing to Mersenne in a letter dated 9 January 1639: "Et ie croy qu'il y continuellement quelques parties de cete matiere subtile qui se ioignent aux cors terrestres, en sorte qu'il n'y a point de matiere en tout l'univers qui ne puisse receuoir successivement toutes les formes" (AT 2:485).

80. Aquinas 1918–30, III, 22. This is an obscure passage about celestial bodies becoming more perfect by acquiring proper places by analogy to matter acquiring a proper form; there Aquinas says: "thus matter receives successively all the forms towards which it is in potential [sic enim successive materia omnes formas suscipit ad quas est in potentia]." See Carraud forthcoming. As Vincent Carraud also points out, Leibniz rails against the proposition on a number of occasions (for example, see his letter to Molanus, ca. 1679, in Leibniz 1989, pp. 240–45).

vince would be another story,[81] but let us look briefly, in conclusion, at some of the ways in which Cartesians, or people who came to be Cartesians, in the seventeenth century used, or rejected, the traditional terminology in their work.

Cartesians

We began by considering how, both within and beyond the scholastic tradition, strict hylomorphism had been challenged before Descartes initiated his dualistic program. What did his followers, from the 1640s to the end of the century, do with the form-matter relation? There is no single answer to that question; let us look at a few cases, which range—though not in chronological order—from more to less (or least) retentive of Aristotelian concepts.

Father René le Bossu believed there was no conflict between the ancient and the modern master; he published a "parallel" of their principles in physics, based largely, admittedly, on the work of Descartes's disciple Jacques Rohault.[82] As he saw the situation, Aristotle had been teaching beginners, and so started with what was obvious to everyone, namely, the sensible things around us, and asked what they were made of: thus, the statue, for example, was made of bronze. For Aristotle, the "first" matter is, le Bossu declares, the proximate matter, of which this thing is made. Descartes, at a more advanced stage of science, considered the matter common to everything, which is extended substance, and every particular is given a form by the way that general matter is shaped. Their principles are not so different. For Aristotle there are three: privation, matter, and form. (Le Bossu offers a unique account of how privation works; it is what is involved when you peel away the form to get at the matter: you deprive the thing of its form!) Descartes, thinking of what constitutes bodies, not of how we know them, needs only two principles: form and matter. Consider also their parallel definitions of matter and form:

> *Aristotle:* Matter is the proper and immediate subject of which each thing is made; it makes us know the form which is drawn from it. Extension is of its essence, but is not its essence.
> *Descartes:* Matter is an extended substance, which is the common subject of which all material things are made.
> *Aristotle:* Form is what makes each thing be what it is, and makes us know it in particular.

---

81. For that story, see Chapters 8 and 9.
82. Bossu 1981. See also Grene 1993.

> *Descartes:* The material form is an arrangement of parts of matter; it makes each thing in particular be what it is.
> *Aristotle:* Matter and form are two equally substantial parts of the composite: each being an incomplete substance in purely material beings, composed of body and material soul.
> *Descartes:* Matter and form are not equally substantial, but matter alone is a substance.[83]

At least that is some concession of difference, but in the main it is the same form-matter universe, considered from different stages of the development of scientific knowledge.

Another way to reconcile Descartes and Aristotle was to contrast Aristotle himself (who really was an incipient Cartesian) with the wicked commentators and Scholastics who corrupted him. That was how Johannes de Raey managed it: you take Aristotelian passages out of context and read them as a Cartesian.[84]

If they were giving formal academic lectures, however, even would-be Cartesians had to put their arguments into a standard scholastic, that is, more or less Aristotelian, framework. This is characteristic both of Heereboord in Leyden and of Jean-Robert Chouet, who introduced Cartesianism into the academy at Geneva.[85] Heereboord remarked in a letter that, after the Boots, nobody really could justify talk about "forms,"[86] and in his disputed questions, collected under the title *Meletemata,* he made heavy use of Pemble's arguments, arguing defensively that no one could consider that Fellow of Magdalen other than devout and decent.[87] But he could not organize his lectures in so shocking a manner. Chouet had been taught by Gaspard Wyss at Geneva and David Derodon at Nimes, both of whom questioned the doctrine of substantial forms; Derodon was even an atomist. Yet in their formal lectures both Heereboord and Chouet had to follow the traditional format, working their Cartesian, or other innovative, views into the context of the usual scholastic headings, including prominently matter and form.

Cartesians presenting their views outside the academy were not so constrained; yet even they made concessions to form and matter as the principles of physics. In Rohault's *Traité de la physique,* for example, there are two such chapters. In chapter 6 Rohault introduces form and matter—minus

---

83. Bossu 1981, 284–87.
84. See Grene 1993, 73–77.
85. Heereboord 1668; for Chouet, see Heyd 1982, esp. chap. 4, pp. 116–44.
86. See the letter to Colvius of 8 April 1642, excerpted in AT 8B:197. See also Verbeek 1992 and Ruler 1995.
87. Heereboord 1680, 162.

privation—as "the principles of natural things," and, after some chapters on divisibility, motion and rest, and other happily material matters, he inserts one on form (chap. 18) before proceeding to the elements.[88] It is a rather apologetic chapter, dismissing substantial forms in nature but allowing for some "essential" forms, such as water's liquidity, as against "accidental" forms, such as water's coolness. However, these notions seem to do little work in the treatise as a whole. Somewhat similarly, later in the century, the Cartesian Pierre-Sylvain Régis, in his *Whole System of Philosophy*, gets around to some brief chapters on form—by which he means simply shape—and from there goes straight to vortices.[89] Presumably there were among their readers still enough individuals schooled in scholastic thought, so that they would find pleasant a small amount of this kind of discourse; there seems no other reason for it. It certainly does very little work.[90]

In a work such as Antoine Le Grand's *Entire Body of Philosophy according to the Principles of the Famous Renate Descartes*, finally, the situation is again a little different. This was a very popular work, first published in Latin in 1671. The English edition of 1694 seems to be the seventeenth-century equivalent of a coffee-table book: illustrated "with more than an hundred sculptures" and "Endeavoured to be so done, that it may be of Use and Delight to the Ingenious of Both Sexes."[91] Le Grand's only chapter on form is entitled "There are no Substantial Forms really distinct from Bodies."[92] Forms do enter into his account of matter, but in a thoroughly Cartesian manner. Here he concedes some agreement between Cartesians and Peripatetics, but he turns the "agreement" in a Cartesian direction, much as de Raey did. The Aristotelians' *"First Matter,"* he declares, is "nothing but an inadequate conception of a *Body*, as it may be conceived by us without any *Figure, Hardness, Softness, Colour*, or any other Modifications, and only as Extended,

---

88. Rohault 1987, 1:21–22, 102–5.
89. Régis 1970, La physique, bk. 2, chap. 1, "Des Formes des Estres purement materiels en general"; chap. 2, "Examen du 1. Chapitre du 1. Livre de l'ame, dans lequel on pretend qu'Aristote etablit les formes substantielles"; chap. 3, "De la division des Formes materielles en general"; pp. 390–97.
90. We should note that Régis, in his preface, goes out of his way to assert the compatibility of the Cartesian doctrine of matter and form with the creation story. God created matter in an instant and subsequently informed it: "Nous supposerons donc que la Matiere a esté créée en in instant, mais que l'ornement, c'est à dire, que la disposition, l'ordre et l'arrangement de divers corps qui se sont formez par l'application successive de ses parties les unes les autres, s'est fait successivement en la maniere que nous allons décrire, ou en quelque autre maniere équivalente" (Régis 1970, 388–89 [unnumbered]).
91. Le Grand 1972, title page. For more information about Le Grand, see R. A. Watson, introduction to Le Grand 1972; see also Grene 1993, 80–81.
92. Le Grand 1972, 102.

and consisting of Three *Dimensions*." From this there follow, he continues, four propositions on which both schools agree:

> The *First Proposition:* The First matter is *without form:* For in this, the Notion of *Extension* is abstracted from all Modification, that belong to the *Essence* of a *Body.*
>
> The *Second:* The *Matter* of all *things* is the *same;* for all *Bodies* agree in this first or Primary *Attribute* of a *Body, viz. Extension,* neither is there any distinction between them with relation to the *Matter.*
>
> The *Third:* Every *thing* may be made of every *thing;* or, according to the *Peripateticks, Matter* is capable of all *Forms:* For since all *Bodies* agree in *Extension,* all their differences must be from their various Modifications; If therefore there be an *Efficient Cause* sufficient to alter these *Modes,* it may make every thing of Every thing. On this *Proposition* are grounded all those *varieties* which are observed in *Bodies.*
>
> The *Fourth:* A *Body* as such, or the First *Matter,* is *Ingenerable and Incorruptible.*[93]

This is truly a very Cartesian Aristotle; indeed, the third proposition sounds very like Descartes's echo of a Thomistic thesis in *Principles* III, article 47. Thus Le Grand seems to be presenting a purer Cartesian natural philosophy than Rohault or even Régis do. Although he occasionally lapses into more Aristotelian discourse,[94] there is no hint of hylomorphic thinking anywhere.

These are a few examples of both pre- and post-Cartesian discussion of form and matter. As we mentioned earlier, scholastic approaches also persisted throughout the century. However fateful the ultimate outcome for the hylomorphic perspective, the way to it was complicated and sometimes devious. By Descartes's time, matter was already struggling for independence of the form for which it used to long. Descartes endorsed that independence by assigning extension to it as its principal, indeed, its only essential, attribute.[95] At the same time he intensified the transformation of form into mere shape or size, and of course he separated the mind—the only immaterial finite substance—from any but a temporary task of informing the body. Some form-matter talk lingered on, even among his followers, but, except for the persistence of a minority scholastic position, it was as good as over.[96]

---

93. Le Grand 1972, 94.
94. Le Grand 1972, 322.
95. But see also Des Chene 1995b for scholastic precedents for that move.
96. Scholastics (and members of religious orders, in this case, Jesuits and Minims) such as Honoré Fabri (1666) and Emanuel Maignan (1653) discarded substantial forms. There were even Schoolmen, such as Casimir de Toulouse (1674), who tried to merge Gassendi's atoms with Peripatetic philosophy. We should not, of course, forget the Leibnizian reestablishment of matter-form.

[5]

Scholastics and the New Astronomy on the Substance of the Heavens

The traditional view in the history of science is that the astronomical observations made by Galileo circa 1610 (following Copernicus's mathematical speculations) precipitated the scientific revolution in the seventeenth century. The new astronomy thus required a new physics. There is much to be said for this view, but it should also be noted that it does not make much sense of either the Scholastics' or Descartes's activities: if the new astronomical observations were so decisive for the new science, then the Scholastics were simply irrational in maintaining the old science, and Descartes's project was marginal at best for the new one. Descartes does not seem to be driven by the new astronomy; on the contrary, his physics appears to flow from metaphysical-theological contemplations. As early as 1630, Descartes wrote to Mersenne telling him that for the past for nine months he had worked on nothing other than his metaphysics. He had been trying to establish the conditions for discovering the foundations of physics: knowledge of God and of the self. "I think that all those to whom God has given the use of reason have an obligation to use it principally so as to know him and to know themselves. That is the path I tried to take when I began my studies; and I can say that I would not have been able to discover the foundations of physics had I not looked for them along that road."[1] Although we do not have the text of this "lost metaphysics," we can be fairly sure that it would not have been much different from the later metaphysics we do know. Given that Descartes thinks of the self as the only form, he then thinks of matter as extension, differentiated by degrees of motion and size. His radical mechanistic explanations and his

---

1. AT 1:144.

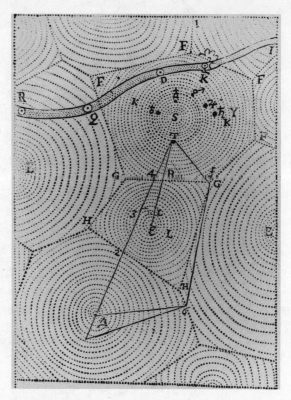

Our Neighborhood according to Descartes (1632)

(astonishing) astronomical speculations are an outgrowth of these principles; as a consequence we get matter distinguished only quantitatively, motion as change of place, vortices, etc. Even such phenomena as sunspots are explained according to Descartes's parsimonious metaphysics. In *Principles* III, articles 94–119, Descartes accounts for sunspots with an analogy of scum bubbling up to the surface of a liquid. Novas are then stars whose sunspots become so dense as to be concealed from our view; planets (such as the earth) and comets are stars whose spots have hardened into a crust—the latter moving with high speed and the former slowly, thus remaining at a fixed distance from the center of its vortex. Descartes's project, then, while it is not inconsistent with the expected general pattern, does not fit it very well.

According to the traditional view, furthermore, debates of the Scholastics are simply dismissed. Here is a typical pronouncement:

> The arguments brought forth against [Galileo's] new discoveries were so silly that it is hard for the modern mind to take them seriously. Galileo did not

bother to reply to them in print, though he often answered many of them in his personal correspondence with his friends, often quite amusingly.... One after another, all attempts to cleanse the heavens of new celestial bodies came to grief. Philosophers had come up against a set of facts which their theories were unable to explain. The more persistent and determined adversaries of Galileo had to give up arguing and to resort to threats.[2]

Thus, Schoolmen were bookish philosophers who failed to grasp some obvious facts, constructed silly arguments, and ultimately resorted to threats. For confirmation of this view, one can always point to the story about Cesare Cremonini, who refused to look through Galileo's telescope. That ostrichlike story does capture the imagination. It is made worse because of Cremonini's position as chief philosopher at Padua in 1610 and "the leading Aristotelian in Italy, and perhaps in Europe" until 1631.[3]

The most recent book on the scientific revolution begins with an account of Galileo's discoveries, concentrating on sunspots. "Sometime between the end of 1610 and the middle of 1611 the Italian mathematician and natural philosopher Galileo Galilei (1564–1641) trained the newly invented telescope on the sun and observed dark spots, apparently on its surface."[4] The author distinguishes between Galileo's observations and his conclusions and correctly points out that the Scholastics had an explanation for the sunspots: "But whereas other contemporary observers reckoned that the spots were small planets orbiting the sun at some considerable distance from it, Galileo was sure, based on calculations in mathematical optics, that they were 'not at all distant from its surface, but rather contiguous to it or separate by an interval so small as to be quite imperceptible.' "[5] But the author rejoins the tradition when he discusses the significance of Galileo's interpretation of the sunspots:

> Not Galileo's observations of sunspots but his particular interpretation of those spots was widely taken as a serious challenge to the whole edifice of traditional natural philosophy as it has been handed down from Aristotle (384–322 BC) and modified by the Scholastic philosophers of the Middles Ages and Renaissance. Galileo's views on sunspots, along with a body of other observations and theorizing, profoundly questioned a fundamental Aristotelian distinction between the physics of the heaven and that of the earth. Orthodox thinking, from antiquity to Galileo's time, had it that the physical nature and principles

---

2. Drake 1957, 73–74.
3. Drake 1978, 162, 446.
4. Shapin 1996, 15. I surely hope that Galileo did not look at the sun through his telescope but rather projected its image, using a *camera obscura* technique.
5. Shapin 1996, 15. The "calculations in mathematical optics" must refer to the spots' lack of observable parallax.

of heavenly bodies differed in character from those obtained on earth.... the sun, the stars, and the planets obeyed quite different physical principles. In their domain there was no change and no imperfection.... These are the reasons orthodox thinking located comets either in the earth's atmosphere or at least below the moon: these irregularly moving ephemeral bodies were just the sort of thing that could not belong to the heavens.[6]

This account, while making more sense of the Scholastics than previous writers had, still does not get the story right, and thus fails to understand the powerful appeal of scholastic principles and the range of their explanations, or the importance of the Cartesian maneuvers.

Another story might help. Less than a year after Cremonini's refusal to look into Galileo's telescope, on 4 June 1611, the Jesuit College of La Flèche held the first memorial celebration of the death of Henry IV. Henry le Grand, the patron of La Flèche, had his heart sent to be buried at La Flèche the previous year, with all appropriate pomp. The students, including even the young René Descartes, composed and performed verses in French and Latin for the memorial. The compositions were published for posterity as *Lacrymae Collegii Flexiensis,* or *The Tears of the College of La Flèche* (1611). One of the French poems from the collection had the unlikely title "Concerning the death of King Henry the Great and on the Discovery of Some New Planets or Wandering Stars Around Jupiter, Noted the Previous Year by Galileo, Famous Mathematician of the Grand Duc of Florence."[7] In the poem the sun revolved around the earth; but, taking pity on the sorrow of the French people for the loss of their king, it offered them a new torch: the new stars around Jupiter. Hence the sonnet combined a pre-Copernican view of the sun with an announcement of Galileo's discovery the previous year of the moons of Jupiter. The news of Galileo's telescopic observations obviously traveled very fast; the students of La Flèche, and by inference, their Jesuit teachers, seem to have had no objection to the telescope. They even praised its use and Galileo's results. To be sure, it also seems that the students might not have been made aware of the significance of the observations, that is, their possible use as evidence for the Copernican point of view.

It is well known, but perhaps not well enough appreciated, that the Jesuit mathematicians of the Collegio Romano, including Christopher Clavius, accepted most of Galileo's astronomical observations. In 1611, Cardinal Bellarmine wrote to the Jesuit mathematicians asking whether they could

6. Shapin 1996, 15–17.
7. *Sur la mort du roy Henry le Grand et sur la descouverte de quelques nouvelles planettes ou estoilles errantes autour de Jupiter, faicte l'année d'icelle par Galilée, célèbre mathématicien du grand duc de Florence,* in Rochemonteix 1899, 1:147n–148n. See Chapter 1.

confirm Galileo's telescopic observations.[8] The mathematicians responded quickly in the affirmative, confirming that more stars can be seen than ever before, there are "handles" of Saturn, phases of Venus, and moons around Jupiter. However, they did not think that mountains on the moon can be observed using the telescope. They granted the great inequality of the moon's surface but added that "Father Clavius thinks it more probable that the surface is not uneven, but rather that the lunar body is not of uniform density and has rarer and denser parts."[9] As Clavius's reluctance to admit it showed, the existence of mountains on the moon was treated as a conclusion, not as a direct observation. Briefly, the argument was that in the context of scholastic theories of the transmission of solar light by the moon, the appearance of mountains on the moon could have been caused by the rarefaction and condensation of the lunar matter, resulting in the differential transmission of the solar light, and not by the alleged rough surface of the moon reflecting the solar light.[10] I will return to this issue.

The stories depicted so far do not favor the image of the ostrich. Rather, the Aristotelians in the seventeenth century seem to have been in the position of people who have seen their favorite theories belied by observations. The question before them was, Could small modifications save their theories or did they have to make major overhauls? The Aristotelians' story is not often told from that perspective, a perspective in which late Scholasticism reacts to celestial novelties, makes adjustments to its theories, that is, changes and survives. It is the story I wish to tell, using the writings of seventeenth-century Schoolmen, followers of Aristotle, at Paris and at Jesuit and non-Jesuit colleges around Paris. The topics I will treat involve celestial matter, especially lunar and solar spots, and comets.

But first, it should be pointed out that from this perspective, unless the Collegio Romano Jesuits truly did not understand the implications of what they were admitting, one can infer that Galileo's observations could be directly accommodated within Aristotelian cosmology. This is not an insignificant result, since at least one of these observations, that is, the phases of Venus, is often given as a definitive observation for Copernicanism.[11] A little-known representation of the world clearly confirms the above result. There is a schematic drawing of the whole universe in a mathematical textbook written by Jacques du Chevreul, an Aristotelian who taught mathematics and physics at Paris in the 1620s.[12] Du Chevreul's picture is a representation of a three-dimensional (or solid) eccentric-epicycle geocentric

---

8. Galileo 1890–1901, 11:87–88.
9. Galileo 1890–1901, 11:92–93.
10. See Ariew 1984.
11. See Ariew 1987b.
12. Chevreul 1623, 257.

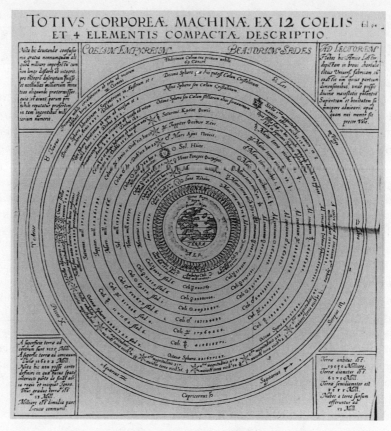

The Universe according to Eustachius a Sancto Paulo (1609)

model, in the general tradition of Aristotle and Ptolemy. It can be readily shown that the solid eccentric-epicycle model was very fashionable during the first half of the seventeenth century. Christopher Clavius, who taught an entire generation of Jesuit teachers of mathematics, used the model in his *Sphaera*.[13] The model lasted well into the second half of the seventeenth century; a nice picture of a solid eccentric-epicycle for the system Sun-Mercury-Venus can be found in Jean-Baptiste de la Grange, *Les principes de la philosophie* (1682).[14] Moreover, it was also the model represented in Eustachius a Sancto Paulo's immensely popular *Summa philosophica quadripartita*, a Paris textbook published in 1609 and republished about 20 times until 1648.[15] Thus, du Chevreul's overall structure was faithful to the pre-

13. See Lattis 1989.
14. Grange 1682.
15. Eustachius a Sancto Paulo 1609, 2:96. Eustachius's diagram can be found in Clavius's earlier *Sphaera*, but as a theory of the sun. See Lattis 1989, 384.

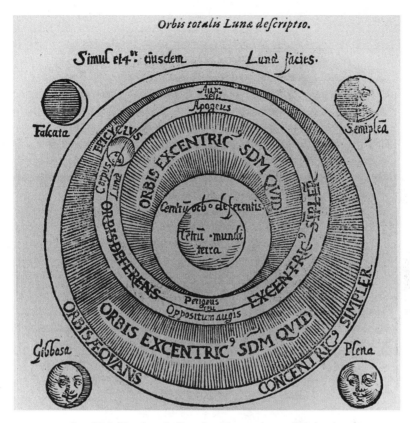

Our Neighborhood, Showing Eccentrics and Epicycles,
according to Eustachius a Sancto Paulo (1609)

Galilean Aristotelian image, that is, he depicted a geocentric three-dimensional epicyclic-eccentric spherical universe. But du Chevreul's representation of the world also made the minimal corrections required by Galileo's observations. In response to the phases of Venus, du Chevreul had Venus and Mercury revolving around the sun; moreover, Jupiter was given four moons and Saturn was given two (consistently with the Jesuit account of Saturn). None of du Chevreul's modifications seem to have required any significant changes in the traditional view.

Sunspots and Moonspots (the Man on the Moon)

We can also add that du Chevreul had "sunspots" revolving around the sun. Sunspots were not discussed in the exchange between Bellarmine and the Collegio Romano mathematicians. But they seem to have caused no fundamental problem for the Schoolmen, having been widely interpreted, in

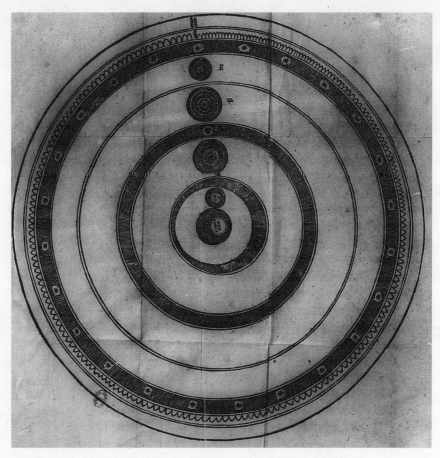

The Universe according to Jacques du Chevreul (1623)

France, as small planets going around the sun. From du Chevreul to the Provencal astronomer Jean Tarde,[16] to the Jesuit mathematician Pierre Bourdin,[17] to Descartes's acquaintance Charles Sorel,[18] sunspots were thought of as small planets, like Mercury. According to the accepted theory, that would simply make them denser parts of celestial spheres around the sun. As the Dominican Antoine Goudin put it, "It could be asked what are the spots that obscure the sun. I've already said, they seem to be a denser part of the celestial spheres which, when nearing the sun, following unknown revolutions, become interposed between us and the star and obscure its brilliance with their passage; Mercury has been taken for a sunspot in this way for a long time."[19] However, even if the sunspots had not been thought of as small planets rotating around the sun, but as genuine spots on the sun itself, they still would have caused few problems, since they could have been treated in the manner of moon spots, that is, denser parts of the celestial body itself. As René de Ceriziers explained in *Le philosophe français:* "We can gather from this small discourse that if there were spots on the sun or stars similar to the spots on the moon, they would arise only from the diversity of their parts"—by which he meant their density and rarity.[20]

There is a full exposition of the problem in Goudin's *Philosophie,* an influential late-seventeenth-century textbook which had at least a dozen editions and even a definitive nineteenth-century Latin edition. The problem, according to Goudin, is as follows: "Spots are produced and disappear in the sun itself; they have sometimes been seen in such quantity that the star became obscured by them, something which lasted a whole year.... And now, as recently observed, spots are produced on that star, less sensible no doubt, but more numerous and similar to smoke or to a dark fog, and this cannot be explained without substantial alteration of the star."[21] But Goudin, as an Aristotelian, cannot accept substantial alteration in the heavens. So the problem is particularly difficult; he first rehearses the traditional answer but must then reject it:

---

16. See Baumgartner 1987.
17. "La figure F represente certains astres qui tournent autour du soleil et que d'aucuns appellent, les taches du soleil, ainsi qu'il se dira cy-après" (Bourdin 1661, 128).
18. "L'on a decouvert qu'il y en avait [des astres] au dessous du soleil, lesquels faisoient paraistre en luy comme des taches. Il est vray que l'on a remarqué qu'ils faisoient leur revolution tout autour, tellement qu'ils prennent quelquefois le dessus. Il y peut en avoir encore en beaucoups d'autres endroits que les yeux des hommes ne peuvent descouvrir: mais tant des uns que des autres, il semble que l'on ne scauroit rapporter la vraye cause de leur rang, puisque nous ne pensons pas qu'ils se cedent rien l'un à l'autre dans leur situation" (Sorel 1634, 69).
19. Goudin 1864, 85. Goudin also confesses: "Ce sujet est, du reste, fort peu connu; personellement je n'en ai pu rien voir, parce que, depuis plusieurs années, les taches n'ont paru que tres rarement."
20. Ceriziers 1643, 194.
21. Goudin 1864, 64.

Assuredly, the argument derived from the motion of the sunspots is not unworthy; there are sunspots disappearing when advancing from the borders to the middle of the disk; others appearing instantaneously in the middle; and then another having been seen alone expanding into several. Perhaps the causes of these phenomena are in the bodies not on the sun or in the upper heaven, but in the sublunary air; perhaps also they are condensed exhalations that, in some way, follow the motion of the star. . . . But this hypothesis cannot explain the spots observed recently, since it has been practically demonstrated that they are very near the sun.

Goudin argues instead for a version of the doctrine that sunspots are small stars around the sun:

We must therefore say that these spots are denser portions of the celestial spheres near the sun, which, following a determined path, are encountered with the sun and sometime reunite and sometime separate, then show us one, then another of their faces. Allowing us to see them at times, then disappearing; they first seem to us a single spot, then several separated spots, and these apparent spots grow or diminish according to the combinations of these spheres, all of this being nothing more than a certain number of optical effects. . . . One easily conceives that the heavens proximate to the sun, such as those of Venus or Mercury or others nearer still, can have certain parts that are more opaque, and that when these parts meet with the sun, they can show it to us covered by spots, in the way that less experienced eyes saw Mercury. But the doctrine that the sun lets some smoke, fog, scum, ashes, or other such things escape—and all that in order to explain the spots seen there and the fact that these productions remain so long attached to that luminous star—is something I cannot understand.[22]

For Goudin it is difficult to account for the spots, given that they appear and disappear. He explains them as optical illusions, combinations of denser portions of the celestial spheres in motion near the sun; he tells us that Mercury had once been mistaken as a sunspot. He seeks an explanation, a piecemeal adjustment of his Aristotelian theory, but he rejects any explanation involving substantial change in the heavens, which would be a denial of the fundamentals of his Aristotelian cosmology. This explains why the theory of sunspots as small stellar objects lasted so long; as late as 1705, one can find in the *Philosophia universalis* of Jean Duhamel: "It is probable that the sunspots are nothing other than small planets revolving around the sun."[23]

As de Ceriziers and Goudin show, the problem of sunspots can be reduced to the problem of lunar spots. And the theory of lunar spots was one

---

22. Goudin 1864, 64–65.
23. Duhamel 1705, 5:45.

of the theories that seventeenth-century Scholastics did not think needed much change. Following Clavius's suggestion, du Chevreul handled the lunar spots as rarefactions and condensations of the celestial matter,[24] as did Parisian textbook writers during the first half of the seventeenth century (that is, all of the ones I have consulted). The doctrine lasted well into the second half of the seventeenth century. One can find in the *Cursus philosophicus* of the Jesuit Vincent that "the spots are the parts of the moon that are denser or rarer."[25] One can even find the doctrine at the start of the eighteenth century in Jean Duhamel's *Philosophia universalis:* "It is probable that the spots of the moon derive their origin from the inequality of density and rarity of the parts of the moon."[26] The account they all accepted predated Galileo's 1609–11 observations, as the following passage from the most popular French-language textbook of the seventeenth century, Scipion Dupleix's *La physique* (1603), made clear: "Aristotle's commentators, after having made a precise inquiry, almost all defend Averroës' opinion that the moon has some parts which are thicker than others, and that they receive more light from the sun than the thinner parts to the extent of their thickness (for by itself the moon is opaque and dark). It then happens that we see clearly some of the parts and not the others. This resolution seems to me the best, given that none more appropriate has been found."[27]

The one early French textbook writer, Théophraste Bouju, who in 1614 held that there is no sphere of fire and no absolute division between the sublunary and superlunary world—who even referred to the telescope as evidence for his point of view—continued to maintain the standard theory for the propagation of light.[28] His general theory was that heaven and the stars do not have their own light, but receive the light of the sun differentially on account their density and rarity.[29] His lunar theory followed from the general theory; he accounts for the lunar spots by the "fact that light does not

24. Chevreul 1623, 166–68.
25. Vincent 1660, 3:319; see also Pierre du Moulin 1644, II, chaps. 6 and 7: "Les estoiles tant fixes qu'errantes ne sont autre chose qu'une partie du ciel plus espaisse que les autres parties. Car le reste du ciel est diaphane et transparent: Mais les estoilles par leur epaisseur arrestent la lumiere du soleil comme miroers, et nous la renvoyent. . . . Quant aux taches qui paraissent en la lune, elles ne sont autre chose que parties du corps de la lune moins espaisses que le reste, et qui n'arrestent point les rayons du soleil. Comme quand un miroer en certains endroits n'a point d'argent vif en derriere."
26. Duhamel 1705, 5:46.
27. Dupleix 1990, 442.
28. Bouju 1614, 1:405–8 (chap. 18, "Que l'element pretendu du feu n'est point"). Bouju claims to be giving an exposition of all of philosophy: "Le tout par demonstration et auctorité d'Aristote, avec eclaircissement de sa doctrine parluy-mesme."
29. Bouju 1614, 1:374–75 (chap. 23, "De la lueur du soleil, et de la lumiere qu'il communique aux autres corps"): "Tellement que tout le corps du ciel et les estoiles n'ont aucune clarté qui mérite de nom de lumière que par le seul bénéfice du Soleil: non toutefois en une mesme manière, ains diversement: car les autres la reçoivent comme un vase ou un mirrouer,

reflect against the parts on the moon which are rarer than others, given that these parts lack the thickness to stop and retain the light." In this way, Bouju could maintain a fairly standard theory for the spots on the lunar surface: "[The moon appears] with this variety and deformity in it which is not in the other stars, as the Dutch lenses clearly show. And the cause of this defect might be that the moon is close to the lower bodies, in which obscurity and deformity dominate."[30] Lest anyone accuse him of giving up too much Aristotelian doctrine, Bouju was careful to uphold the (de facto) incorruptibility of heaven: "Since it does not appear to us . . . that the sun is of another matter than the other lower bodies, its incorruptibility must arise from its more excellent form than theirs or because contrary agents which can corrupt and alter it do not rise up to it, although it is corruptible with respect to its nature, in the manner of air and other elements."[31]

As with sunspots, the lunar spots are explained without having to resort to substantial change in the heavens. However, Bouju, possibly under some Stoic influence, had accepted a version of Aristotelianism in which there can be substantial change in the heavens, at least in principle; for most Aristotelians, substantial change in the heavens would not have been acceptable. Fortunately, even if one admitted that the spots on the moon could only have been accounted for by postulating mountains, one did not have to accede to there being substantial change on the moon.

That is a point Goudin wanted to make. Unlike many other Aristotelians, Goudin accepted the conclusion that the spots were to be interpreted as mountains on the moon: "This opaque substance, which is properly lunar substance and different from the essentially lucid celestial substance, even though it has the form of a globe, is not however perfectly smooth on its surface; it has some parts which are depressed and others which are elevated, as if they were valleys, canyons, mountains, and hills."[32] But he prepared an explanation whereby the moon overall would be polished and spherical, with a transparent substance in between the seeming mountains and valleys:

> Since absolute void is not possible, we must believe that this unequal surface is equalized by a transparent substance that allows everything to be seen and

---

duquel à cause de son espoisseur, elle est réfléchie et renvoyée: et eux en deviennent lucides: parce que leur parties estant espoisses, elle ne les peut pénétrer. Et quant au reste du ciel, elle y est receue comme en un moyen transparent, qui la porte, duquel elle n'est point réfléchie, à cause de sa transparence et rareté: comme il se connoit la nuict que le ciel n'est veu que par la clarté des estoilles."

30. Bouju 1614, 1:388–89 (chap. 42: "De la lune").
31. Bouju 1614, 1:380–81 (chap. 31, "Comment le ciel peut estre et n'estre pas incorruptible").
32. Goudin 1864, vol. 3, "De la nature de la lune, et sa face semée de taches," p. 96.

everything to be illuminated by the rays of the sun. No telescope has been able to tell us if this substance is fluid or if it is solid in accordance with the Peripatetic idea. But nothing prevents us from considering the globe as spherical in the sense that the inequality of its surface would be corrected by distinct and solid parts, in truth, but ones that are completely diaphanous and fill the voids left over by the opaque part.[33]

Still, after due deliberation, Goudin accepted the conclusion that the spots on the lunar surface are best explained as shadows caused by mountains on the moon: "These fleeting spots that vary according to the position of the sun are shadows projected by the parts that jut over the less elevated ones."[34] All the while, however, Goudin understood that the mere existence of mountains on the moon did not require the postulation of substantial change in the heavens, and thus did not require the abandonment of Aristotelian cosmology:

> The proof of the incorruptibility of the lunar body and of its difference from the elements that surround it results from all these phenomena. Let us prove it. Even though we see a variety of mountains and valleys on the body of the moon, we do not, however, see any alteration, any change, apart from the changes produced by shadow and light; now, if some alteration were produced there as with us, we should be able to see it, since we see the change resulting from shadow and light so clearly that, for a given point, with a strong enough telescope, we can observe in half an hour the decreases and increases of light and shade. Therefore, there is only a succession of light and shade on the lunar globe; thus the lunar globe is a body whose essence is different than ours.[35]

One can conclude that Galileo's astronomical observations seem to have posed no serious problem for the Parisian textbook writers of the seventeenth century. The Parisians accepted the observations made by the telescope. They made the modifications of their theories required by the observations. They changed their theories of Venus and Mercury. They added stars around Jupiter and Saturn. They attempted to accommodate the observations of sunspots and moon spots by treating sunspots as stellar objects, by explaining them as rarefactions and condensations of the celestial matter around the sun, and by reducing them to the same status as that of lunar spots. Finally, they grappled with the issue of lunar spots by accepting them not as anomalies that needed a new explanation but rather as new phenomena requiring only a slight change in Aristotelian cosmology.

33. Goudin 1864, 3:97.
34. Goudin 1864, 3:98.
35. Goudin 1864, 3:99–100.

Comets

We have not yet discussed another astronomical observation often given for the demise of the Aristotelian traditional view—comets. Comets seem to provide a powerful argument against the heterogeneity of the sublunary and supralunary regions of the world.[36] The standard view is that two astronomical observations—the appearance of a new star (nova) in 1572 and Tycho Brahe's measurement of the parallax of the comet of 1577, from which he concluded that the comet was in the heavens—had dealt a heavy blow to the traditional view of the immutability and perfection of the heavens. This view was appreciated at the time, as can be seen in a student thesis from circa 1622. In a series of propositions about Copernicus and the new astronomy, the author discusses the observational consequences of the Tychonic system, though without naming Tycho. Among the consequences is the assertion that "comets and new stars would be generated and would move in the heavens above the moon."[37] But Tycho Brahe's parallax measurement was neither universally accepted nor without conceptual difficulties. As Dupleix explained,

> Since comets are elevated very high into the region of air and are moved and shaken by the celestial bodies that carry them, the elementary fire, and the upper air, and also because they look like true stars, because of their flame, several ancient philosophers, and even Seneca and the common people ignorant of this matter still, take comets to be true stars. But this ignorance is too crass, given that stars are all in the heavens and comets are in the region of air below the moon, as is demonstrated by astronomical instruments.[38]

Regiomontanus, *de Cometis*

The text is accompanied by a marginal note: "Regiomontanus, *de Cometis.*" The note indicates that Dupleix accepted Regiomontanus's parallax measurements, not Tycho's.

The dispute might have been complicated by the exchange between Galileo and Horatio Grassi about the parallax of the comet of 1618. Grassi, a Jesuit astronomer, wanted to argue against Aristotle based on the lack of observable parallax for the comet.[39] Galileo disputed his findings, arguing that one cannot use the parallax of a comet to calculate its location. Parallax is a valid method only when one has a real and permanent object;[40] for

---

36. This topic has recently been discussed in Barker and Goldstein 1988.
37. "Sententia Copernici de motu terrae circa solem omnes apparentias non saluat, & habet alia incommoda. Si terra constituatur centrum circulorum quos luna, sol, & stellae fixae conficiunt, sol vero eorum quos reliqui planetae, facile omnia defendentur. Mars nonnunquam terris propior quam sol apparet" (Borbonius circa 1622).
38. Dupleix 1990, 423–24.
39. See Drake and O'Malley 1960.
40. Galileo, *Il saggiatore*, in Drake and O'Malley 1960, 186–87.

example, one cannot use the parallax of a rainbow to calculate its location. Thus, the parallax of a comet (or its lack of parallax) cannot give us its supralunary location and is not evidence for concluding that the Aristotelians are wrong (or for concluding further that there is an imperfect terrestrial object in the heavens) unless, of course, we had previously accepted comets as objects whose nature is terrestrial, and not meteorological phenomena or mere appearances. Galileo in 1623 proposed that comets are luminous reflections of atmospheric exhalations, an account similar to the one he had proposed in 1606 and similar to the Aristotelian account. Here is a standard account of comets from Bouju's *Corps de philosophie* in 1614.

> The fire of which the comets are enflamed and of which they burn is slow and moderate; comets are not raised up on account of the weight of their matter, but they move from east to west in accordance with the motion of heaven, although they do not do so with regularity. The height of their motion is less than that of the planets and other stars; it demonstrates that they remain in the middle region of the air, in the same way as do those lights in the form of stars which seem to fall from heaven, which are only meteors, of the nature of comets, and not true stars, being generated and corrupted almost in the same instant.[41]

Parisian textbooks from the period do not seem to have integrated the debate between Grassi and Galileo into their discussions, except perhaps to the extent that they seem genuinely undecided about whether comets are sublunary or supralunary. Du Chevreul discussed the method of parallax in general in his *Sphere*,[42] he then discussed the issue of parallax and comets. Du Chevreul was obviously an Aristotelian, but he did not seem to need to decide the question fully.[43] In his lecture notes on Aristotle's *Meteorology* a year later, he still claimed that comets are sublunary flames.[44] Similarly, François Le Rées, in his *Cursus philosophicus,* had a long discussion of

---

41. Bouju 1614, 1:600–601 (I, Phys. 11, chap. 12, "Des comettes"). See also Marandé 1642, 331–32: "La comete se forme d'une exhalaison chaude, seiche, grasse et onctueuse, en la plus haute region de l'air, de maniere que venant à se circuler par l'espace de quelques jours, elle se sublime peu à peu, deschargeant tousjours de ce qu'elle a de plus aqueux, vaporeaux ou grossier; et comme elle se refere de plus en plus en soy mesme par ce rejet, battue qu'elle est de la chaleur du soleil qui contribue à la sublimer et la purifier; elle parvient enfin à tel point d'onctuosité visqueuse et vaporeuse, qu'elle faict corps et s'allume aux rayons du soleil qu'elle repousse fortement et dont la percussion luy fait prendre feu, qu'elle conserve apres autant que la matiere luy dure; que si elle en peut attirer d'autre semblable par sa chaleur, elle l'a convertit en soy mesme, et dure en cet estat jusques à ce que son propre feu l'ait entierement devorée."
42. Chevreul 1623, 47–51.
43. See Chevreul 1623, 83–85.
44. Chevreul 1624, fol. 547.

comets and parallax, pro and con Aristotle, including Tycho's observations, finally concluding for Aristotle—barely. He argued that new stars are comets. He did not resolve the question of parallax, but merely discussed various options he seemed to think were all ultimately consistent with Aristotelian principles, if not Aristotle's actual doctrine about comets being fiery exhalations. In order of preference, the possibilities were: comets could lack parallax and be above the moon, not in the upper region of air; they could be old stars; or their parallax could be observed, consistently with Aristotle's opinion that they are sublunary fiery exhalations.[45] Le Rées's possibilities had been previously discussed by Jean Crassot in his 1618 *Physica*.[46] Thus Parisian indecision about comets preceded the debate between Galileo and Grassi. It continued for a considerable time, as evidenced by Louis de Lesclache's 1651 textbook. Under the topic of "fires which form mainly in the higher region of air, like comets, and which happen only infrequently," Lesclache wrote, "The difficulties that philosophers have in discovering the place of comets must not occupy the minds of those who seek knowledge of natural things in order to acquire knowledge of God."[47] Others took a more conservative view. René de Ceriziers discussed various opinions concerning comets, including the possibility that comets are engendered in the heavens but are corruptible, that they are exhalations attracted by the sun, and that they are wandering stars having different motions above and below the heavens than the planets (requiring the hypothesis of fluid heavens, which he rejected).[48] De Ceriziers asked, "But why would we not see comets ordinarily, if they were stars? Why would they not have the shape of other stars?" and in 1643 he concluded, "Let us believe with the Philosopher that comets are exhalations that are ignited in the upper region of air."[49]

The issue might have been given a decisive turn by another textbook writer, the Protestant, Pierre du Moulin, in 1644: "Aristotle holds that comets are fiery exhalations; but the astronomers of this time have observed that a comet was above the moon. If that comet was a fiery exhalation, it would have always kept its tail behind it, in the manner of a torch, which when carried always keeps its flame behind it. And the fact that it was

---

45. Le Rées 1642, vol. 2, part 2, pp. 14–26.
46. Crassot 1618, 475–76.
47. Lesclache 1651, 68, table 11 of the Physics.
48. Ceriziers 1643, 363–64: "Pour les comètes, certains ont cru qu'elles s'engendroient dans le ciel, ils avouent par la corruptible: d'autres qu'elles se formoient des exhalaisons que le soleil attiroit jusques-là. Appolonius tient que ce sont des astres vagabonds, qui se montrent et se retirent à divers temps, ayant leur mouvement du haut du ciel en bas à la différence des autres planettes qui tournent, ce qui suppose les cieux liquides."
49. Ceriziers 1643, 363–64.

seen by so many in so many countries demonstrates its great height."[50] Du Moulin's paragraph was unusual in several respects. He gave a nice argument that some comets are not fiery exhalations, based on anti-solarity. The tails of some comets have been observed to point away from the sun, and not away from their direction of motion, as might have been expected if comets were fiery exhalations. And he constructed an argument about the distance of comets based on their being visible at many places at the same time, a commonsense way of getting the parallax arguments across. Du Moulin was not original in this respect, as the following passage from Bouju—about a nova, not a comet—demonstrates. As we have previously indicated, for Aristotelians novas are the same kind of entities as comets, and pose the same kind of problems.

> We have seen in our time, during 1572, a new star appearing in Calliope and lasting two years. In the beginning this star seemed to surpass Venus in size and clarity and two months later it decreased in these respects, such that it no longer seemed to exceed a star of the third magnitude; it kept this quantity for the duration of two years, when it disappeared. It cannot be said that this star was in the air where comets usually happen, because it appeared in the same way to all who saw it, in whatever region it was, and it always moved from east to west like the other stars; this could not happen if it were located only in the middle region of air, the place of comets.[51]

Du Moulin concluded that there are two kinds of comets, sublunary fiery exhalations à la Aristotle and celestial miraculous objects (which, since they were miraculous, Aristotelian physics did not have to explain): "I believe that both opinions are true and that there are two kinds of comets. The comets of the first kind are miraculous and celestial and above the moon; and consequently they are more meaningful."[52] By the second half of the seventeenth century, one can find Aristotelians accepting comets as celestial objects, as indicated by the following student theses from Jesuit colleges: "Both the form and matter of comets are celestial; thus a comet is a star, not a fire"; and "Comets are celestial; in truth they are planets."[53] One

---

50. Moulin 1644, IV, chap. 3, pp. 103–4. By 1644 du Moulin no longer resided at Paris (which he did from 1599 to 1620) but at Sedan.
51. Bouju 1614, 1:381.
52. Du Moulin 1644, IV, chap. 3, p. 104. Brockliss (1987: 375) argues that dismissal of anomalous physical phenomena as miraculous was a standard move at the time. See also Théophraste Renaudot's conférence on comets, in which three out of four speakers are concerned with the signification of comets, their portending the death of a great personage. There is a discussion of some of the conférences (including the one on comets) in chap. 2 of Sutton 1995, 19–40, but the author seems to consider the views represented in the conférences as heretical.
53. Bonaviolensis 1657; and Vigne 1666. One can find a similar theory in de la Vigne's thesis: "II. Coeli tres numero et specie distincti; figura rotuindi sunt; natura corruptibiles, si

[114] CONTEXT

can even find the successor to the two comets theory, in which both kinds of comets are nonmiraculous:

> It must be said that there seem to be two kinds of comets: some are permanent bodies placed in heaven, appearing and disappearing with respect to us; others are only meteors produced by terrestrial exhalations, appearing in the highest regions of air and being ignited there. Proof of the first part. Most of the comets recently observed are certainly higher up than the moon. Now, there cannot be any new production in this part of heaven, as needed for the second opinion. Therefore, these are permanent bodies.[54]

As all the Parisian textbook writers seem to have indicated in their own ways, there was no difficulty with comets being stars, except that if they were stars, they could not have become sublunary. As far as I can tell, no one ever suggested (nor could they have lived to suggest) that a comet crossed the division between the sublunary and supralunary world. On the other hand, if a comet, seen as a star, had a path that carried it across the celestial spheres, then a revision of the solid eccentric-epicycle model would be called for. One might be led to adopt a Tychonic or semi-Tychonic system on account of comets, and this path was taken by many Jesuits. We are given such an account in a public thesis by Jean Tournemine, a student at the Jesuit College of La Flèche. Tournemine sketches a theory of three heavens, supported by "Apostolic authority," as he says. Of the three heavens, two are traditional—the empyrean and the firmament—but the third, the planetary heaven, has a fluid substance, "as shown by astronomical observations."[55] This theory of the heavens is clearly at odds with Aristotelian principles; the hypothesis of a fluid first heaven (and indeed the theory as a whole) seems more suitable for the Tychonic scheme.[56]

Empyreum excipiamus: liquidum praeterea Firmamentum. Eorum materia est eadem sublunariu."

54. Goudin 1864, 3:317.

55. Joannes Tournemyne (La Flèche, 1642), as edited in Rochemonteix 1889, 4:365–68. It is interesting to note that "Apostolic authority" is invoked for the theory. Cf. Bellarmine 1984; also Grant 1987.

56. As late as 1651 one can find Paris writers denying the Tychonic system and defending what they called a semi-Copernican system (the earth rotating on its own axis, but not revolving around the sun—in other words, a return to the speculation of Nicole Oresme). See Garnier 1651. For similar kinds of arguments, using observations to conclude for fluid heavens, see Schofield 1981. See also Bourdin 1646, a single volume in which two small cosmological treatises are bound together: *Sol flamma* and *Aphorismi analogici*. In these works, Bourdin argued that the sun is a blazing fire, a position inconsistent with the Aristotelian theory of the heavens, as Bourdin knew quite well (pp. 1–3: "auctores, et argumenta sententia negantis [Aristoteles]") and supported by such innovators as Descartes. He even referred to Descartes as someone who holds the position: "novissime a Renatus des Cartes solem docet esse flammam" (p. 5). Bourdin's basic argument is that the sun is a body on which there are sunspots and small torches, as the telescope rendered evident. Thus the sun is corruptible matter, not

But the Tychonic system was also taken up as a modification of a general Aristotelian point of view. Still, an Aristotelian would prefer the hypothesis of solid heavens, as Goudin amply demonstrated:

> It seems more probable that the heavens are solid. First objection. The solidity of the heavens cannot be accounted for given the facts observed recently. The improvements of the telescope and the serious studies of our astronomers have made this hypothesis incapable of being sustained. For example, we notice that Mars appears at times higher and at times lower than the sun; that Venus and Mercury revolve around the sun and are at times below it, at times above it, and at times to its side; that there are satellites around Jupiter et de Saturn; that the sun and Jupiter rotate on their axes, etc.
>
> Reply. Saint Thomas tells us to refer to the experts with respect to such questions; if the phenomena observed by the astronomers really do seem in opposition to the solidity of the heavens, we would no doubt abandon our conclusion; but in the midst of so many people who yell so loudly, we are still allowed to listen to some very renowned astronomers, among whom is Giovanni-Domenico Cassini, Director of the Royal Observatory, eminent light of astronomical science, and these astronomers tell us that, until now, none of the observed phenomena are contradicted by the hypothesis of solid heavens.[57]

But Goudin, like many others in the second half of the seventeenth century, was ultimately able to accept the hypothesis of fluid heavens: "The heavens can be fluid and continue to be incorruptible. It is not impossible for a fluid body to be incorruptible: the air, water, and blood of the Blessed after the Resurrection, as well as their vital and animal spirits, will be fluid, in the same way that ours are now; yet they will be incorruptible."[58]

All the features of our discussion of comets within a late Aristotelian context, suitably modified, can be recovered in a pamphlet on the comet of 1664-65, written by Jacques Grandamy, a Jesuit teaching at the Collège

---

incorruptible ether as Aristotle had it: "sol est corpus; in quo sunt eiusmodi maculae, et faculae, ut patet ex telescopio, et parallaxi, quae docet haec omnia non distare a sole; ergo sol est corruptibilis," pp. 7–9; "atqui sol paret flamma (ut patet rescipiendi per telescopium; quo, ut docet Scheiner lib. 2 *Rosa Ursina*, cap. 4. deprehenduntur in sole multa flammae signa)," pp. 14–16. In the *Aphorismi analogici*, such considerations compelled Bourdin to adopt a Tychonic cosmology. He moved from an explanation of sunspots on analogy with foam bubbling up from the sea, to there being three regions of stars and planets, to magnetic phenomena affecting both the earth and the heavens ("Explicantur maculae solis exemplo spumarum maris," pp. 44–46; "Distinguuntur stellae et planetae in tres partes seu regiones," pp. 49–50; "De influxu magnetico mundi tum caelesti tum terrestri," pp. 50–52; "De terminis fluxus magnetici mundi," pp. 52–53). He rejected the Copernican hypothesis, however, claiming that the earth stays still ("Terra quies probatur primo," pp. 65–66).

57. Goudin 1864, 3:68–69.
58. Goudin 1864, 3:63.

de Clermont in Paris. Grandamy argued that comets, being located above the moon, have to be of the same kind as the stars and other celestial bodies; and while he acknowledged the Aristotelian tradition of comets as sublunary exhalations, he did not give much credence to it: "The matter of this comet is celestial, the same as that of planets and stars, since it is as celestial as they are, having been born in the heavens and having its motion there, as we will show in what follows. However, at times, some have seen comets in the air and lower than the moon, as some have wanted to assert, but which I do not guarantee."[59] The problem remained how to distinguish the seemingly corruptible comet from the incorruptible celestial bodies:

> But I cannot and must not give to the comet in question here, which is born and resides in the heaven, any matter other than the one which it has in common with the stars and the planets, which likewise have their domain in the celestial region, with this difference, however, that the fixed and wandering stars have been made from the beginning of the world from a celestial matter which was liquid and fluid, and have received from their Author a proper consistence in order to eternalize their duration and to receive light better and to reflect it more clearly. Instead, comets are made with the same celestial matter somewhat condensed, as needed to reflect the light it receives and to transmit it partially. This celestial light is not so rarefied that it cannot send to us the light it receives nor so thick that the rays of the sun cannot penetrate it.[60]

Grandamy's choice of cosmology, his liquid and fluid heavens, was obviously Tychonic, something he indicated by referring to the authority of modern authors (and to the Church Fathers, as we have already seen):

> I assume here according to the ancient opinion of the saintly Church Fathers and the most certain philosophy of the modern authors, that the whole mass of the celestial machine between the firmament and the heaven of the moon is composed of a liquid and fluid matter, which is easy to move and clear and transparent; that comets and other lesser bodies whether old or newly formed appear in this matter, and that they are moved as the fish in the ocean or the birds in the air. This appears manifest by the birth, the course, and the loss of comets and even of certain stars, which have sometimes appeared anew and disappeared afterwards; this allows us to judge that several other bodies are formed in the same manner, have their course, and disappear in this vast region of the heavens, without our perceiving it. The four satellites of Jupiter, which were unknown for all antiquity, and which only the use of the telescope

---

59. Grandamy 1665, chap. 1, "De la matiere de la Comete," p. 1.
60. Grandamy 1665, chap. 1, pp. 1–2.

have allowed us to know in our days, sufficiently show this truth by their errant motion.[61]

Grandamy's adoption of a Tychonic cosmology should not prevent us from seeing that he was most keen to place his explanation of comets and his fluid heavens into a general Aristotelian framework. Having explained comets as celestial objects, Grandamy then explained that the changes comets underwent were not substantial changes but accidental ones, in keeping with the Aristotelian thought that there is no generation or corruption of any substance in the heavens:

> Besides the birth and the destruction of these comets, neither the condensation nor the rarefaction of the celestial matter prove that there occurs any generation or corruption of a substance, but only an accidental alteration, as the one occurring between flowing and iced water, or between milk and blood and clotted milk and blood, and finally between soft juices and liquids and the same things when they have hardened. For it is the same substantial form which is in the whole mass of the heaven, and in each of its parts, solid as well as fluid; and the substance of the stars is not different from that of the rest of heaven, in the same way that knots in wood do not have a different matter from the rest of the wood, and metals and diamonds and pearls have the same matter and the same substantial form before and after their being hardened.[62]

Grandamy's Aristotelian conclusion was reiterated in his second chapter, "De la forme de la Comete": "Therefore, in this way our comet has the same matter and the same substantial form as the rest of heaven and all of its parts, as have the fixed and wandering stars. In this way it is only an accidental form that distinguishes them from the planets, and this form consists in that it is composed of a head and of a tail and that it has a motion which is proper to it."[63]

In fact, Grandamy kept the standard view of the transmission of sunlight, that it depended upon whether the celestial body transmitting the light was more or less rarefied; he even applied the account to the tail of comets: "The tail is a work of light and a rough and imperfect image of the sun; for the sun, like an excellent painter, paints its light in as many places as it has or can bring its rays. All the rays and all the species of light are as many images of the sun, more or less perfect, according to the diversity of the bodies in which they are encountered, which are mirrors representing their objects in different ways according to their diverse shapes and according to

---

61. Grandamy 1665, chap. 1, p. 2.
62. Grandamy 1665, chap. 1, p. 2.
63. Grandamy 1665, chap. 2, pp. 2–3.

whether they are more or less polished or more dense or more rarefied."[64] Grandamy used his "optical" theory of comets to refute any opponent who might have argued that comets are fiery exhalations above the sphere of the moon, in the heavens:

> In fact, there are three reasons that prove manifestly that the tail of the comet is only the effect of the light of the sun penetrating the head of these same comets and illuminating the heavens beyond it from behind. For first the tail of the comet is always opposed to the sun in a direct line, the head always being between the sun and the tail of the comet on the same line. . . . In this way one can refute the opinion of those who believe that there is a fire in the heavens which forms the matter of the exhalations of comets and their tails. . . . This opinion does not explain why, in many cases, the tails of comets are constantly opposed to the sun, nor why often they turn in an instant from east to west.[65]

One can find Grandamy's general theory as late as 1705 in Jean Duhamel's *Philosophia universalis*. In his first conclusion Duhamel asserted that it is probable that some comets are exhalations in the highest region of air,[66] but in his third conclusion he also asserted that some comets can be celestial globes that reflected the sun's rays by means of their density.[67] In his fourth conclusion he asserted that some comets can be wandering stars that appear and disappear.[68] In his last conclusion he speculated that the mechanism for the formation of these celestial comets was that they were formed from collections of wandering stars which were insensible and which had previously evaded our senses.[69] Duhamel's mechanism for the formation of comets was different from Grandamy's, but the basic intuition was the same, that comets would be explained as changeable celestial phenomena, but not as the corruption and generation of celestial substance. Duhamel had previously argued that those comets higher than the moon can have been generated and corrupted in appearance and with respect to us but not physically and with respect to themselves. They were collections of stars; thus the generation of comets was nothing more than the aggregation of stars, and their corruption was nothing other than their separa-

---

64. Grandamy 1665, chap. 2, p. 3.
65. Grandamy 1665, chap. 2, pp. 4–5.
66. Duhamel 1705, 5:138.
67. Duhamel 1705, 5:139.
68. Duhamel 1705, 5:139.
69. Duhamel 1705, 5:140.

tion.⁷⁰ The Aristotelian theory of comets, suitably modified, seems to have been extremely resilient.⁷¹

We have seen the Aristotelians make modifications of their cherished theories in order to account for Galileo's celestial observations. But, it could be said, although the Aristotelians did not look quite like ostriches refusing to see the obvious, for a very long time they did maintain some false theories about solar and lunar spots and comets. True, but it is not obvious that they should have done otherwise. The evidence for radical change was not clear without hindsight. As we have already pointed out, even Galileo maintained a roughly Aristotelian theory of comets as sublunary events. Moreover, Descartes, who did break from the traditional view of comets, placing them in the heavens and accepting their generation and corruption, also held a false view of them as very hard, fast moving, massive bodies.⁷²

Taken in a piecemeal fashion, the Aristotelian system appears to have been rich enough to provide explanations for the various astronomical novelties: Aristotelian science in the seventeenth century was much like an organism, living and adapting. Descartes provided another complete system to compete with it;⁷³ soon Gassendi and Hobbes would do the same, and a battle for supremacy would be waged among the various creatures.

---

70. Duhamel 1705, 5:31. Ironically, the Scholastics' principles seem to require them to explain celestial novelties by means of a mechanistic philosophy; thus the revolution seems ultimately to be an application of celestial mechanic to the physical world, rather than, as it is usually thought of, the application of terrestrial physics to the celestial world.

71. These conclusions are likely to be confirmed for other portions of the Aristotelian theory under attack and in other sociopolitical contexts. See Reif 1962 and 1969; also Brockliss 1995a.

72. Descartes, *Le Monde*, chap. 9; *Principia Philosophia* III, art. 119–22.

73. On the need to provide complete explanations and not to try to explain phenomena piecemeal—that is, concerning Descartes's criticism of Galileo's scientific methodology—see Ariew 1986.

PART II

Debate and Reception

[6]

Descartes, Basso, and Toletus: Three Kinds of Corpuscularians

There is a general consensus that one of the more important changes in seventeenth-century philosophy is the movement from what is variously called Scholasticism, naturalism, or animism (held by the Aristotelians) to what is variously called the mechanical philosophy, corpuscularianism, or atomism (held by the moderns): "of central importance to the history of the physical sciences in the seventeenth century and beyond was the revival of the ancient atomistic doctrines of Democritus, Epicurus, and Lucretius."[1] This revival is understood as the attempt to explain the characteristic behavior of bodies in terms of the size, shape, and motion of the small particles that make them up; it is usually accompanied with the elimination of sensory qualities such as heat and cold, color, and taste. Given that atomism was rejected by Aristotle, the emergence of the mechanical philosophy appears to entail the rejection of scholastic philosophy. Historians and philosophers do seem to agree on these points: "Throughout the scientific circles of western Europe during the first half of the 17th century we can observe what appears to be a spontaneous movement toward a mechanical conception of nature in reaction to Renaissance naturalism. Suggested in Galileo and Kepler, it assumed full proportions in the writing of such men as Mersenne, Gassendi, and Hobbes, not to mention less well known philosophers. [One can add Descartes, Boyle, and Newton to the list]."[2] Even the latest history of the

---

1. Garber 1992, 117.
2. Westfall 1971, 31–32. In chapter 2 of his book Westfall argues (from a Newtonian perspective) that the conjunction of mechanism and corpuscularianism was detrimental to the scientific revolution, which needed to detach corpuscular matter theory from mechanical or mathematical theory of motion.

"scientific revolution" accepts this phenomenon and its accompanying demarcation into two camps, the Aristotelians and the moderns, in order to argue that "the superior intelligibility, and therefore the explanatory power, of the mechanical philosophy was more limited than its proponents claimed. Adherents' *conviction* that mechanical accounts were globally superior to alternatives, and more intelligible, has to be explained in historical rather than abstractly philosophical terms."[3] Indeed, it has been argued that the central philosophical and methodological problems of early modern philosophy were not posed by Galileo's mechanics or Copernican astronomy but by the corpuscular or mechanical philosophy: "One of the most persistent, and philosophically disturbing, features of most sciences of the 17th century was the radical observational inaccessibility of the entities postulated by their theories. As numerous scholars have shown, it was the epistemological features of this type of theory which occasioned much of the philosophizing of the 17th and 18th centuries."[4]

Now, it happens that Descartes would not have fully agreed with these characterizations. In *Principles* IV, article 202, "The philosophy of Democritus differs from my own just as much as it does from the standard view of Aristotle and others,"[5] Descartes explains the relations between his philosophy and those of Aristotle and Democritus in a symmetrical fashion. In the preceding article, he had already claimed that there are particles in each body that are so small they cannot be perceived by the senses. That claim is the only point of agreement between his philosophy and that of Democritus: "It is true that Democritus also imagined certain small bodies having various sizes, shapes and motions, and supposed that all bodies that can be perceived by the senses arose from the conglomeration and mutual interaction of these corpuscles." Surprisingly, Descartes also presents the claim as a point of agreement between his philosophy as that of Aristotle. Descartes first puts aside the possible objection that the rejection of Democritus's philosophy might have been based on the fact that it "deals with certain particles so minute as to elude the senses, and assigns various sizes, shapes and motions to them." According to him, "no one can doubt that there are in fact many such particles"; thus, he gives four other reasons for the rejection of Democritean atomism: (1) Democritus supposed his corpuscles to be indivisible; (2) he imagined a vacuum around the corpuscles;

---

3. Shapin 1996, 57
4. Laudan 1981, 23, "A Revisionist Note on the Methodological Significance of Galilean Mechanics."
5. Of course, the *Principles* has to be put into context as a teaching text, a work that Descartes hoped might be used in the schools; for that reason, Descartes surely minimized the differences between himself and Aristotle and maximized those between himself and Democritus. See Ariew 1996.

(3) he attributed gravity to these corpuscles; (4) either he did not show how things arose only from the interaction of corpuscles or his explanations were not entirely consistent. Descartes affirms the first three reasons as valid: he himself rejects the indivisibility of corpuscles, demonstrates the impossibility of vacuum, and argues that there is no such thing as gravity in any body taken on its own. He then leaves to others the question of the fertility and consistency of his results. He ends the article with the pronouncement that he rejects "all of Democritus' suppositions, with this one exception [of the consideration of shapes, sizes and motions]," and that he also rejects "practically all the suppositions of the other philosophers." Thus, it is clear to Descartes that his method of philosophizing has no more affinity with the Democritean method than with any of the others. Again, the only relation that holds between his philosophy and Democritus's, according to Descartes, is "the consideration of shapes, sizes and motions." But Descartes also asserts that he shares this consideration with Aristotle and all other philosophers: "as for the consideration of shapes, sizes and motions, this is something that has been adopted not only by Democritus but also by Aristotle and all the other philosophers." Thus, according to Descartes, the important differences are then the actual principles he uses—his "suppositions"—not any basic epistemological differences dealing with mechanism as opposed to naturalism.

The sharp separation between the scholastic and mechanical or corpuscularian philosophy can be challenged on other grounds. The second half of the seventeenth century also saw the rise of "Peripatetic atomism," which, according to the general consensus, is an oxymoron. But that is the title of a philosophy text by Casimir of Toulouse;[6] it is also the attitude adopted by the Jesuit Honoré Fabri and others who introduce corpuscularian principles and explanations into scholastic philosophy.[7]

Instead of demarcating between the Scholastics and the moderns, it would better to take Descartes's lead and to investigate intellectual relations holding between various philosophers. It might turn out that we would affirm Descartes's judgment that his philosophy differs as much from atomism as it does from Scholasticism; or better, turning the judgment around, we might discover that Descartes's philosophy has as much or more in common with Scholasticism than it does with atomism. Thus, I propose to look at the contrasts Descartes draws between his philosophy and that of the early atomist Sebastian Basso, "one of the most influential

---

6. See Casimir de Toulouse 1674.
7. See Blum forthcoming. Daniel Sennert might also be thought to be an early scholastic corpuscularian. See also Thorndyke 1941–58, vol. 7, chap. 23, for a discussion of Etienne Natalis, another Aristotelian corpuscularian, and Fabri.

authors among the early corpuscularians,"[8] and the one he might have drawn with those of contemporary scholastics such as Franciscus Toletus.

Mersenne and Descartes on Basso

Basso's reputation seems assured by the publication of his book *Philosophiae naturalis adversus Aristotelem*,[9] which was cited by his more famous contemporaries, Mersenne and Descartes among others.[10] As early as 1624, Mersenne ranks him among the atheists because of the book. In *L'impiété des Deistes*, after having discussed such "despicable" authors as Charron, Cardano, Machiavelli, Bruno, the "accursed" Vanini, "and similar rogues," Mersenne talks about the work he is writing against them: "I do not want to spend much time on this subject, since I expect to refute everything these authors stated so inappropriately in the Encyclopedia I am preparing in the defense of all truths and against all sorts of lies, in which I will examine more diligently what has been advanced by Gorlaeus, Charpentier, Basso, Hill, Campanella, Bruno, Vanini, and a few others."[11] Mersenne proceeds to gives examples of the "impertinence" of these authors. He complains specifically about the adherence of Charpentier and Gorlaeus to the principle that "all things are made and derived from nothing" and that of Gorlaeus and Hill to atomism, that is, to Hill's "Epicureanism" and to the doctrine "that inside bodies there are atoms which have quantity and figure." Mersenne has nothing substantial to say about Basso, other than having placed him in such disreputable company; according to him, "ultimately, they are all heretics, which is why we should not be surprised that they agree, being all as thick as thieves." A year later, in *La vérité des sciences*, he ranks Basso again in roughly the same company, but this time it is as an anti-Aristotelian: "Franciscus Patrizi has tried to discredit Aristotle's philosophy, but he made no more progress than Basso, Gorlaeus, Bodin, Charpentier, Hill, Olive, and several others, who raise monuments to Aristotle's fame through their writing, since they are not able to strive high enough to bring down the flight and glory of the Peripatetic Philosopher, for he tran-

---

8. Meinel 1988, 182.
9. Basso 1649 [1621]. For more details about Basso's life and times, see Lüthy 1997.
10. Also Isaac Beeckman, Joannes Magnenus, the Boates, and others. Frey 1628, chap. 17, is entitled "Villonii theses et cum ipso Clavius, Garassus, et Bassonis, cribantur." Frey does discuss Garasse, Villon, and de Claves, and in previous chapters he discusses Ramus, Campanella, Gassendi, Telesio, Patrizi, Bacon, and others as failed critics of Aristotle, but, unfortunately, he forgets to say anything about Basso.
11. Mersenne 1624, 237–38. We should note that Bruno and Vavini were burned at the stake and that Campanella was imprisoned for more than two decade by the Catholic Church; we should also note that the "atheists" and "rogues" were not all atomists.

scends everything of the senses and imagination, and they grovel on the ground like little worms: Aristotle is an eagle in philosophy and the others are like small chicks who wish to fly before they have wings."[12] The two early Mersenne references to Basso do not engender much confidence in the claim that Mersenne read Basso—that he read anything more than the title of the work. However, there are also indications in Mersenne's correspondence that he was acquainted with some of the contents of Basso's work. Indeed, he discussed certain of his doctrines with Descartes.

Basso was reasonably well known to Descartes, who, like Mersenne, listed him with Bruno and Vanini, among others. In the only place where Descartes mentions Basso by name, he ranks him among the *novatores*—the innovators: "Plato says one thing, Aristotle another, Epicurus another, Telesio, Campanella, Bruno, Basso, Vanini, and all the innovators (*novatores*) all say something different. Of all these people, I ask you, who is it who has anything to teach me, or indeed anyone who loves wisdom?"[13] But this passage is not the only place in which Descartes gives his opinion of Basso's philosophy. Descartes had written to Mersenne on 8 October 1629: "As for rarefaction, I am in agreement with this physician and have now made up my mind about all the foundations of philosophy; but perhaps I do not explain the *aether* as he does."[14] According to the editors of Descartes's works, the person in question was perhaps Villiers, a physician from Sens,[15] but this identification was corrected as a reference to Basso by the editors of Mersenne's correspondence.[16] Thus, with Descartes, we finally have specific comments about the contents of Basso's philosophy.

---

12. Mersenne 1625, 109–10. Mersenne's point of view has found modern proponents. In *A History of Magic and Experimental Science*, Lynn Thorndyke asserts:

> The publication in Geneva in 1621 by Sebastian Basso of twelve books of natural philosophy against Aristotle testifies as much to the abiding influence and even dominance of the Stagirite as it does to the existence or success of opposition to his teachings.... Professedly at least Basso's own treatise is not a step forward to modern science but a step backward to the natural science of Aristotle's predecessors. Actually he deals with roughly the same set of topics and problems as do the Aristotelian works of natural philosophy and their subsequent commentators. And while professing to "restore the abstruse natural science of the ancients," Basso cites and depends upon his immediate predecessors like Zabarella and Piccolomini a great deal. Or he finds it advisable to refute and attack Scaliger, Toletus, and the Conimbricences. In other words, his book ... is more in the nature of a critical commentary upon the Aristotelian philosophy as developed during the medieval and sixteenth centuries than it is a new departure in the direction of modern science." (Thondyke 1941–1958, 6: 386–87)

13. Letter to Beeckman, 17 October 1630, AT 1:158.
14. AT 1:25.
15. AT 1:30.
16. Mersenne 1933–88, 2:302, 307–8; cf. the appendix of the 2d revised edition, AT 1:665.

Another probable reference to Basso is contained in an addition to Descartes's correspondence, the letter to Huygens of 28 March 1636. A few days before the letter, on the twenty-third, Huygens had offered Basso's *Philosophiae naturalis adversus Aristotelem* to Barlaeus, who refused it, since he already owned a copy. In the letter itself, Descartes thanks Huygens for a book he just sent him: "The book which you did the favor of sending me yesterday is truly a recent blessing, for which I very humbly thank you. I do not know whether I dare say this to you, since you have had the patience to read it, but I am persuaded that my reveries will not be insufferable to you, for, if I remember, it is only good for destroying the opinions of Aristotle, and I seek only to establish something, so simple and manifest, that the opinions of all others would agree with it."[17]

Assuming, what is probable, that Descartes was speaking of Basso in all these passages, what do they tell us about Descartes and Basso? We must first note that there was a change of mind in Descartes about Basso. Initially, in October 1629, Descartes agreed with him about rarefaction, but disagreed about the ether. Then, a year later, we have the identification with the *novatores* in the context of an ill-tempered letter to Beeckman concerning what anyone can teach another. Basso does not have anything to teach Descartes, any more than anyone else (unless he can convince him by his reasons). Finally we have the disavowal of Basso: he is good only for destroying Aristotle's opinion. Descartes denies that he shares this intent, claiming that he seeks only to establish something so simple and evident that everybody would agree with it. To reach a fuller understanding of the relationship between Descartes and Basso, however, we need to recall their doctrines on rarefaction and the ether; and this requires us to discuss their views on corpuscles and the void. But above all, we need to compare these doctrines with those of the Aristotle of the Scholastics at the start of the seventeenth century—with Aristotelians such as Eustachius a Sancto Paulo, Scipion Dupleix, and perhaps the early Daniel Sennert,[18] but especially Toletus and the Conimbricenses, authors constantly cited by Basso, and whom Descartes remembers reading in his youth.[19]

---

17. AT 1:602–3.
18. Eustachius a Sancto Paulo 1629; Dupleix 1990; Sennert 1618 (English trans. 1659); Raconis 1651.
19. Toletus 1589; Conimbricenses 1592. For more complete bibliographic information on these and other such commentaries, see Lohr 1988. For the reference in Descartes's correspondence to his remembering Toletus and the Coimbrans, see his letter to Mersenne, 30 September 1640, AT 3:185.

Some Novelties in Late Scholasticism

For a late (or post-Renaissance) Scholastic, there are two kinds of mutation: substantial (or generation and corruption) and accidental (or motion). Generation and corruption are changes in the substance of a thing: the substance acquires or loses a substantial form. Substantial forms are said to be indivisible, not capable of more or less, and not possessing contraries, and thus they cannot be acquired successively and piecemeal. Motion, in contrast, occurs successively between contraries; motion must pass from one contrary as the term from which (*a quo*) to the other contrary as the term to which (*ad quem*). According to its Aristotelian definition, "the actualizing of what is in potentiality insofar as it is in potentiality,"[20] motion is an imperfect actuality, the actuality of a being whose potentiality is being actualized while it still remains in potency for further actualization. In this terminology, in general, to be in actuality is to "participate in a form"; thus, an actuality is "an accidental or substantial form, in succession or permanently." Moreover, to have potentiality is to have a "principle of acting, or undergoing something."[21] A being moves, then, by virtue of the successive acquisition of qualitative or quantitative forms or of places. For example, water becomes hot by the acquisition of heat, which it has the potential for acquiring. Forms in the categories of quantity, quality, and *ubi* or place, have contraries or positive opposite terms. Thus, true motion is only in those three categories, which entails that there are three kinds of motion: augmentation and diminution (in the category of quantity), alteration (in quality), and local motion (in place). But since a thing cannot both be in actuality and potentiality at the same time with respect to the same form, no object undergoing change can be the active source of its own change or motion; rather, it would have to be moved by an agent already possessing the actuality it itself lacks. Water, for example, cannot be the active cause of its own heating, whereas fire can be the cause of the water's heating, given that fire is actually hot and can turn the water's potentiality for heat from potency to act. That which moves, the agent that introduces a form, must possess the form or actuality; that which is moved has the power or potentiality for receiving the form: "in physical change, all these are found: an agent, a patient . . . , and furthermore an acquired form, and a way or medium by which it is acquired."[22] The thing which moves and the thing

---

20. For example, Dupleix 1990, vol. 3, chap. 4, p. 187.
21. Toletus 1589, III, chap. 1, text. 3: "esse in actu est participare aliquam formam . . . Esse actum est esse aliquam formam accidentalem vel substantialem, successivam, aut permantentes . . . Esse autem potentiam, est esse principium agendi, aut patiendi aliquid."
22. Toletus 1589, III, chap. 3, quaest. 1.

which is moved are therefore not the same, resulting in the principle that everything which is moved is moved by some other thing. Another consequence of the definition of motion is that rest is opposed to motion; it is the privation of motion in the thing which is naturally capable of motion and, inasmuch as motion is made to accomplish rest, it is also said to be the end or perfection of motion. However, living things are moved as well by an internal principle of motion, and the elements, that is, the simple bodies, are carried to their natural places by their forms, which tend to their natural places (the natural place of earth being in the center of the universe, surrounded, in order, by the natural places of water, air, and fire).[23]

An important change in the Aristotelian theory of motion was the adoption of *impetus* theory by the late Scholastics, including Toletus, who (along with Julius Scaliger) was usually cited as the authority in favor of *impetus* by textbook authors in the seventeenth century.[24] These authors, usually Jesuits, credited John Buridan, Albert of Saxony, Scaliger, and Domingo de Soto with the doctrine, thereby giving a sketch of its line of descent. *Impetus* was an attempt to solve a difficulty in the Aristotelian theory of motion: the continued lateral motion of a projectile. Aristotle argued not only that everything in motion is moved by something else, but also that the mover must be in contact with the moved thing. In the case of projectile motion, the only thing in contact with the moving object is the medium through which it moves (usually the air). Aristotle's solution was that the mover of the projectile gives the air immediately surrounding it the power to move the projectile further and that this power is passed on through the medium with the projectile. Scholastics such as Buridan rejected this solution and proposed instead that when a projectile is thrown, the mover transmits an *impetus* to it which then continues to act as an internal cause of its continued motion. Buridan treated the *impetus* as a quality inherent in the moving body, proportional both to the quantity of matter of the moving body and to its speed. He believed *impetus* to be a quasi-permanent quality and, consequently, he inferred that, once the moving body was set in motion, it would tend to continue to move under the direction of the *impetus* until some counteracting cause or resistance intervened.[25]

A second major change in late Aristotelianism was the theory of *minima naturalia*, generally discussed in the context of rarefaction and condensation, or change of quantity. As a rule Aristotle was strongly anti-atomist. He thought the continuum could be divided indefinitely. However, he also ut-

---

23. See also Dupleix 1990, vol. 3, chap. 1–15, pp. 173–228; Sennert 1618, I, chap. 8, pp. 23–29 (1659, I, chap. 9, pp. 43–46).
24. See Raconis 1651, 247 et seq.
25. Buridan 1509, VIII, quaest. 12.

tered the obscure assertion that "neither flesh, bone, nor any such thing can be of indefinite size in the direction either of the greater or of the less."[26] This comment took on a life of its own.[27] By the seventeenth century, the resulting doctrine entailed that there were intrinsic limits of greatness and smallness for every sort of living thing. Moreover, since every natural body has a form which is actually determined, every natural body must have determinate accidents and a limited quantity. But the elements or simple homogeneous bodies have no determinate magnitude of themselves and intrinsically. They might be augmented indefinitely, if there were matter enough, and their division could be continued indefinitely. In contrast, the elements do have an extrinsic limitation in respect to the limitation of prime matter: there may not be enough prime matter to sustain a form, and the amount of prime matter is finite. Moreover, elements cannot be condensed or rarefied, that is, they cannot have their quantity changed indefinitely, without being corrupted. For example, earth cannot be as rarefied as fire, and fire cannot be as condensed as earth; when air is condensed too much, it is turned into water, and water overly rarefied is turned into air.[28] Thus, for a late Scholastic, rarefaction and condensation, that is, augmentation and diminution in quantity, could result in generation and corruption under appropriate circumstances. There is, then, a natural minimum of any given element.

A third major change concerned Aristotle's denial of the void and, specifically, of motion in the void. Aristotle concluded against the atomists that motion is impossible in the void, using an argument deriving from his principles of motion. A body moving by impact moves in proportion to the force exerted on it and in inverse proportion to the resistance of the medium in which it is situated. Since a void would provide no resistance, the body "would move with a speed beyond any ratio"[29]—but such instantaneous movement is impossible. Scholastics attempted to soften this and similar arguments, not so as to accept the existence of the void, but so as to accept its possibility, that is, to argue that God could create a void.[30] As a

26. *Physics* 187b 18–21.
27. For a history of *minima naturalia* from Averroës to Toletus, see Duhem 1913–58, 7:42–54; 1908–13, 2d ser., pp. 11–15. There is also an account in van Melsen 1960, vol. 1, chap. 2, pp. 58–81.
28. Toletus 1589, I, cap. 4, quaest. 10–11; Sennert 1618, I, chap. 5, pp 15–16 (1659, I, chap. 5, pp. 27–29); Raconis 1651, 370–77.
29. *Physics* 215b 21–22. Aristotle also argued that the void is impossible, if it is thought to be a place with nothing in it, that is, a location actually existing apart from any occupying body (*Physics* IV, chap. 6–7).
30. Although attacks on Aristotle's views about the void preceded the condemnations of various propositions in 1277, they gained theological inspiration from them (see Schmitt 1987 for the influence of Philoponus in the views of Toletus and the Coimbrans, among others). Among the relevant condemned propositions were "That God could not move the heavens in

consequence, there were numerous discussions of Aristotle's argument about the impossibility of motion in the void, many of them prompted by an internal criticism of Aristotle's position; in particular, it was noted that, in his system, the heavens have a determined speed of rotation but are not slowed down by the resistance of any medium. If one applied Aristotle's reasoning about the impossibility of motion in the void to the heavens, then the heavens would have to rotate with a speed beyond any ratio. On the other hand, rejecting Aristotle's reasoning might lead one to postulate an internal resistance to motion, thus invalidating the conclusion that a body would move with a speed beyond any ratio in the void (that is, instantaneously).[31] A reading of Aristotle that became standard in the seventeenth century was that he denied motion in the void, in contradiction to other ancients, only because they did not posit any other cause for the duration of motion than the resistance of the medium. According to this reading, Aristotle would agree that motion in the void would not be instantaneous and that vacuums, although not occurring naturally, are not impossible supernaturally. The same conclusion was also reached in disagreement with Aristotle. For example, Toletus understood Aquinas as holding against Aristotle that motion in the void would not be instantaneous, and he supported Aquinas's position.[32] Other textbook writers, such as Dupleix, also denied Aristotle's argument against the impossibility of motion in the void, asserting that the speed of motion would be due not just to the resistance of the medium but also to the weight and shape of the moving body.[33]

In sum: unlike what might have been expected, late Aristotelianism countenances a kind of corpuscularianism. Its theory of motion also includes a feature that looks similar to the principle of inertia. Moreover, voids are no longer completely impossible and motion is possible in the void. However, late Scholasticism also reinforces the account of mutation as a change of form, whether it is the acquisition or loss of a substantial form in generation and corruption, or the successive acquisition or loss of accidental forms or places in motion.

---

a straight line, the reason being that he would then leave a vacuum," and "That he who generates the world in its totality posits a vacuum, because place necessarily precedes that which is generated in it; and so before the generation of the world there would have been a place with nothing in it, which is a vacuum."

31. This conception was developed by Thomas Aquinas, among others. See Duhem 1985, chap. 9.

32. Toletus 1589, IV, quaest. IX. Eustachius a Sancto Paulo agreed, calling motion in the void extremely probable (1629, Physica, tract. 3, disp. 2, quaest. 4–5).

33. Dupleix 1992, vol. 4, chap. 10.

Basso on Atoms, the Ether, the Void, and Rarefaction and Condensation.

The structure of Basso's work is clear.[34] Basso alternates exposition and criticism of Aristotle with exposition of what he calls the philosophy of the ancients, that is, of Aristotle's predecessors, by which he means Plato, Empedocles, and especially Democritus.[35] The philosophy he wishes to recover is atomism. According to Basso, the ultimate constituents of bodies are the minimal particles of matter he calls "atoms." These atoms are smaller than the invisible internal organs of the tiniest animals;[36] they are preexistent, incorruptible, and limited in number.[37] They were created by God at the beginning of time,[38] and, setting aside the possibility of their annihilation by God, they are indestructible.[39] Each atom is homogeneous, a simple body possessing a particular nature that persists in mixtures;[40] when atoms enter into composition, they make up natural minima having their own proper natures.[41] According to Basso, there are four kinds of elementary atoms (other than the ether), coinciding with the four scholastic elements.

For Basso, all mutations—generation and corruption, alteration in quality, and augmentation and diminution in quantity—are explicable at the level of the ultimate constituents of matter. Generation and augmentation in quantity are the gathering together of atoms or clusters of atoms; corruption and diminution in quantity are the dispersing of atoms that were previously united. Alterations in quality result from atoms of one kind being substituted for atoms of another: "Infinitely varied parts can be composed in many ways from these primary particles, which are so different among themselves; it is not difficult to understand that by the subtraction or addition of any particle, or by a variation in the arrangement of parts, some parts can be easily converted into the nature of others."[42] Thus, for Basso, completely new generation is an illusion; what happens instead is

---

34. In this section I am in basic agreement with the recent literature on Basso: Kubbinga 1984, Meinel 1988, Nielsen 1988, and especially Lüthy 1997.

35. I cite the 1649 edition; we should note that the subtitle is *In quibus abstrusa veterum physiologia restauratur, et Aristotelis errores solidis rationibus refelluntur.*

36. Basso 1649, 14–15.

37. Basso 1649, 7–8.

38. Basso 1649, 13: "cum agimus de atomis, censemus eas a Deo creatas, quod fuit praemonendum."

39. Basso 1649, 112–13.

40. Basso 1649, 27; see also the resumé, p. 67: "quod ex primis illis qua constituebant, rerum particulis, ita res omnes componi assererent, ut in composito propriam naturam retinerent."

41. Basso 1649, 23.

42. Basso 1649, 72.

the continuous reorganization of atoms.[43] Basso attempts to disprove the scholastic doctrine that the four elements can assume new substantial forms and thus can be generated from one another.[44] Rebutting this doctrine is particularly important to him since, for Scholastics, it contrasts with the doctrines of Democritus about the incorruptibility of atoms.

Basso introduces the ether as a fifth element into the world of atoms, in part to explain rarefaction and condensation, and in part to explain why atoms move.[45] The ether is material and it consists of atoms.[46] It is far more tenuous than the elementary atoms; it permeates every kind of object insofar as it fills the gaps between the atoms of the four elementary kinds.[47] It is the cause of the motion of atoms and, in this way, the cause of the mutations of bodies.[48] Atoms of the four elementary kinds do not possess motive power; they are put into motion solely by something external, namely, the ether.[49] Basso's concept of causation is proto-mechanistic. He does not consider the possibility that atoms may have their own principle of motion.[50] Given that Basso does not envision a principle of inertia and does not mention the late scholastic account of *impetus* (despite the fact that it is discussed by Toletus), introducing empty space into the universe would have unwelcome consequences. The empty space would disrupt the continuity of the ethereal motion. As a result, Basso rejects the void: nature abhors a vacuum.[51]

As long as he denies the void, Basso cannot have recourse to the atomist explanation of rarefaction and condensation as changing ratios of atoms to void. And since he also repudiates the Aristotelian explanation of the phenomenon as change of qualitative form, he is compelled to use the ether. According to Basso, the phenomenon occurs when "all similar natural min-

---

43. Basso 1649, 9–10: "Hinc, monstrant quomodo et ex corruptione nihil de novo generetur: sed tantum earundem partium quarum facta erat copulatio, fiat resolutio"; pp. 215–16: "in materialibus discrepare videantur, in eo tament concordabant, quod, cum nihil ex nihilo fieri constantes assererent, negarent cujus quam generationem aliud esse, quam principiorum praexistentium diversam compositionem."

44. Basso 1649, 118 et seq.: "Certum est autem, ni detur forma substantialis, non dari mutationem substantivam, qualem isti volunt: sed generationem esse nihil aliud quam quod veteres voluerunt."

45. Basso credits the Stoics for having discovered the ether, or *Spiritus* (1649, 300: "En tibi Stoici clari manifestarunt"), though he also maintains anachronistically that Democritus had atoms moving in the ether, in opposition to Aristotle's report that Democritus defended the existence of the void (305).

46. See, for example, Basso 1649, 220; 382–84.

47. Basso 1649, 304, 306.

48. Basso 1649, 308–9, 387–88.

49. See Basso 1649, 300 et seq.

50. Basso 1649, 8, 10.

51. Basso 1649, 300: "vacuum . . . a quo natura abhorret"; 311: "vacuum, a quo Natura abhorret."

ima are diminished," and this results in the object becoming "condensed, and when their quantity is increased, the object becomes rarefied."[52] For these natural minima to increase, the ether needs to interpose itself between the atoms of the body; in this way the gaps between the elementary atoms grow wider as the volume of the body increases: "by penetrating into the parts of air it separates them from one another, occupying a greater place."[53] Condensation is simply the reverse of this process.[54]

Since, for Basso, the atoms of ether do not penetrate or pass through the elementary atoms but permeate only the interstices existing between the elementary atoms,[55] even when the elementary atoms form a compound, the ether is an external principle of motion. The atoms of ether do not become constituent parts of compound bodies;[56] they thus do not play the role of an internal principle. Moreover, the ether is continuous insofar as any atom of ether is always adjacent to some other ethereal atom and participates in the motion that all ethereal atoms share. It is the link that unifies the various particles. It moves every elementary atom according to its own aptitude.[57] Given that part of the aptitude of an atom is having a proper place, each kind of atom has its proper place. Basso is a geocentrist.[58] Thus, the atoms of earth belong at the center of the universe, surrounded, in order, by the regions of water, air, and fiery atoms.[59] Though the ether is the cause of motion, it is totally inert in itself. It is in constant need of being kept in motion by a higher cause. God is the higher cause on which the ether depends, not only for its motion but also for its directing of the motion of the elementary atoms. Thus the ether both depends on God's continually infusing motive force and directs its motion to its proper ends.[60]

Basso, then, is a kind of Democritean atomist (as opposed to a *minima naturalia* theorist): his atoms are indestructible and do not transmute into one another. However, he is a rare kind of atomist, one who denies the void and fills his universe with an ether that does not combine with other atoms. His atoms provide him with a decidedly nonscholastic theory of change: all change is due to the local motion of atoms. Rarefaction and condensation is not the acquisition or loss of a qualitative form; the ether simply inflates or deflates a body. But Basso's universe is inert. Its primary

---

52. Basso 1649, 293–94.
53. Basso 1649, 300.
54. Basso 1649, 301.
55. Basso 1649, 304 et seq.
56. Basso 1649, 308.
57. Basso 1649, 300–303, 307–9.
58. Basso 1649, 279.
59. Basso 1649, 279–80.
60. Basso 1649, 284 et seq., 307 et seq.

cause of motion is God, who imparts motion to the ether; the ether, in turn, moves the various particles.

Early Descartes (1629–33) on Corpuscles, Rarefaction, and Subtle Matter.

Unlike Basso, Descartes is not an atomist.[61] His matter is indefinitely divisible: "every body can be divided into extremely small parts. . . . It is certain that it [the number of these small parts] is indefinite."[62] But like Basso, Descartes uses atomist modes of explanation; he explains the visible by the invisible—macro-phenomena by reference to micro-phenomena; as Descartes says in *Le Monde,* not only the four qualities called heat, cold, moistness, and dryness "but also all the others (and even all the forms of inanimate bodies) can be explained without the need of supposing for that purpose anything in their matter other than the motion, size, shape, and arrangement of its parts."[63] So, like Basso, Descartes does not need substantial forms and does not explain mutation as change of form, whether substantial or accidental. He finds no "forms" other than the ones he has described quantitatively.[64] He even makes fun of the scholastic definition of motion, complaining that its "words are so obscure" that he is compelled to leave them in Latin because he "cannot interpret them."[65] For Descartes, the only motion is local motion; hence he states: "The philosophers also suppose several motions that they think can be accomplished without any body changing place, such as those they call *motus ad formam, motus ad calorem, motus ad quantitatem* . . . and a thousand others. As for me, I know of none except the one which is easiest to conceive . . . , the motion by which bodies pass from one place to another."[66] God is the cause of motion in Descartes's universe,[67] and, because of his immutability, the first law of motion is "that each individual part of matter always continues to remain in the same state unless collision with others forces it to change that state."[68] If a part of matter has a certain size, it will remain that size, unless something else divides it or changes it in some way; if it is moving, it will continue to move, unless something else retards it; and if it is at rest, it will stay at rest

---

61. For more on the topics of this section, including contrasts with Descartes's later views, see Garber 1992.
62. *Le Monde*, chap. 3, AT 11:12.
63. *Le Monde*, chap. 5, AT 11:26.
64. *Le Monde*, chap. 5, AT 11:26–28.
65. *Le Monde*, chap. 7, AT 11:39.
66. *Le Monde*, chap. 7, AT 11:39–40.
67. *Le Monde*, chap. 7, AT 11:37–38, 46–47.
68. *Le Monde*, chap. 7, AT 11:38.

until something else drives it away. For Descartes, rest is a quality of matter just as motion is; it is not a privation of motion.[69]

Since Descartes, like Basso, is also against the void (though for very different reasons),[70] their explanations of rarefaction and condensation also agree: "The corpuscles which enter a thing during rarefaction and leave during condensation, and which can penetrate the hardest solids, are of the same substance as visible and tangible bodies; but you must not imagine that they are atoms.... Think of them as an extremely fluid and subtle substance filling the pores of other bodies."[71] However, in most respects, Descartes and Basso do differ about the nature and function of that subtle fluid. For Descartes, all the elements are quantitatively differentiated—differentiated by shape, size, and motion—and, therefore, all of them are transmutable into one another.[72] Thus, the ether is no different from any other element. There is, however, an aspect of Descartes's subtle matter that does resonate with what Basso says about his ether. Descartes notes that his first and second elements do not enter into composition with his third element: "But we should note that, even though there are parts of these three elements mixed with one another in all bodies, nonetheless, properly speaking, only those which (because of their size or the difficulty they have in moving) can be ascribed to the third element compose all the bodies we see about us.... We may picture all these bodies as sponges; even though a sponge has a quantity of pores, or small holes, which are always full of air or water or some other liquid, we nonetheless do not think that these liquids enter into its composition."[73] Finally, in contrast with Basso, God is the cause of the motion of all matter for Descartes, not just the cause of the motion of the ether; Descartes's world, unlike Basso's, is not inert.[74] Motion, as Descartes says, is a quality of matter; it is preserved by God in his continual recreation.

The contrast can be made more explicit if one considers motion in the void. Descartes agrees with the late Scholastics and disagrees with Aristotle and Basso in holding that motion is possible in the void. For Descartes, the

---

69. *Le Monde*, chap. 7, AT 11:40.
70. *Le Monde*, chap. 4, AT 11:16–23.
71. Letter to Mersenne, 15 April 1630, AT 1:139–40; see also letter to Mersenne, 25 February 1630, AT 1:119, for the same account, using the analogy of the sponge.
72. As Descartes will say, "matter passes successively through all forms it is capable of assuming (la matiere doit prendre successivement toutes les formes dont elle est capable)," *Principles* III, art. 47; and letter to Mersenne, 9 January 1639, AT 2:485.
73. *Le Monde*, chap. 5, AT 11:30–31.
74. Obviously neither Descartes's world nor Basso's is alive in the sense that they are filled with souls or forms (as Leibniz's world will be). In comparison with such a world, both of these would be deemed inert. For God as the cause of motion in Descartes's philosophy, see Garber 1992, chap. 9, esp. pp. 273–92.

state of a body would not change in the void;[75] for a late Scholastic, an *impetus* would not be corrupted in the void. In contrast, according to Aristotle, motion in the void would be instantaneous and, thus, impossible; and for Basso, a gap in the universe would prevent the ether from exercising its activity.

So far, we can confirm that Descartes's account of rarefaction does agree with Basso's and his account of the ether generally does not. Thus, at the level of theoretical explanation, there is both continuity and discontinuity between Descartes's and Basso's philosophies—as there is between those of Toletus and Descartes. However, when Descartes made his statement comparing his philosophy with Basso's, he also intimated that there were significant discontinuities between the two philosophies at a deeper level, that of the reasons for the theoretical explanation, or of the foundations for the theory. We should recall that Descartes also asserts that he has made up his mind "about all the foundations of philosophy." A few months later, Descartes told Mersenne that for the last nine months he had worked on nothing other than his short treatise on metaphysics, establishing knowledge of God and the self, the conditions for discovering the foundations of physics by this means.[76] It was in this very same period that he wrote his letter to Beeckman denying that Beeckman or any of the innovators had anything to teach him. According to Descartes, the person who can teach him "is the person who can first convince someone by his reasons, or at least his authority."[77] Clearly Basso would not be convincing to Descartes, whether because of his reasoning or because of his authority. The reason Descartes gives for developing his own accounts is that they seem to follow from the

---

75. It should be noted that Descartes's rejection of the void is even sharper than the Scholastics', Descartes looking more like Aristotle than the Aristotelians. In *Principles* II, art. 18, Descartes argued for the impossibility of empty space, both in and out of the world. Thinking of a vessel, its concave shape, and the extension that must be contained in this concavity, he asserted: "it would be as contradictory of us to conceive of a mountain without a valley, as to conceive of this concavity without the extension contained in it, or of this extension without an extended substance." In fact, he decided that if God were to remove the body contained in that vessel and did not allow anything else to take its place, the sides of the vessel would thereby become contiguous. However, even though Descartes thinks the void is impossible, he does not think that motion would be impossible in the void if, *per impossibile*, there were a void. Interestingly, Descartes defined void as "a space filled with matter that neither increases nor diminishes its motion" (letter to Mersenne, 15 November 1638, AT 2:442).

76. Letter to Mersenne, 15 April 1630, AT 1:144: "Or j'estime que tous ceux a qui Dieu a donné l'usage de la raison, sont obligés de l'employer principalement pour tacher a le connoistre, & a se connoistre eus-mesme. C'est par la que i'ay tasché de commencer mes estudes; et ie vous diray que ie n'eusse sceu trouver les fondemans de la Physique, si je ne les eusse cherchés par cete voye. Mais c'est la maniere que j'ay le plus estudiee de toutes, & en laquelle, graces a Dieu, ie me suis aucunement satisfait; au moins pense-je avoir trouvé commant on peut demonstrer les verités metaphysiques, d'une façon qui est plus evidente que les demonstrations de Geometrie."

77. Letter to Beeckman, 17 October 1630, AT 1:158.

foundations of his philosophy. The basic principle of Descartes's metaphysics is the real distinction between mind and body. Body is simply extension. As a result, Descartes sets aside the scholastic apparatus of forms and qualities, with the consequence that all change must be grounded in change of place; the Aristotelian account of motion must therefore make way for a new account of local motion, where motion "is a mode of a thing and not some subsisting thing."[78] This perspective, together with some considerations about God as the ultimate cause of motion, suggest to Descartes his first law of motion or principle of inertia for all extended things. Descartes would not be swayed at all by the kinds of considerations that were important to Basso: the reestablishment of the philosophy of the ancients who were eclipsed by Aristotle, the consequent defense of atomism, and thus the need to reject decisively substantial forms and real generation.

This brings us to a final difference between Descartes and Basso, a difference in tactics or rhetoric. Perhaps because of the condemnation of Galileo in 1633, Descartes was more cautious than Basso about openly criticizing Aristotle. As he said, he did not seek to criticize Aristotle, but only to establish something so simple and manifest that the opinions of all others would agree with it. This is consistent with his own advice to Regius, which he followed for most of his life: "you should refrain from public disputations for some time, and should be extremely careful not to annoy people by harsh words. I should like it best if you never put forward any new opinions, but retained the old ones in name, and merely brought forth new arguments. This is a course of action to which nobody could take exception, and yet those who understood your arguments would spontaneously draw from them the conclusions you had in mind." As part of that advice, Descartes cited himself as an example about what to say concerning the Scholastics' substantial forms; he asked of Regius: "why did you need to reject openly substantial forms and real qualities? Do you not remember that ... I said quite expressly that I did not reject or deny them, but simply found them unnecessary in setting out my explanations?"[79] As one can see, Descartes was not the kind of person who, like Gassendi, would have written a book titled *Exercitationes paradoxicae adversus Aristoteleos* or even, like Basso, *Philosophiae naturalis adversus Aristotelem*.

---

78. As Descartes will say in *Principles* II, art. 25: "ac esse duntaxat ejus modum, non rem aliquam subsistentem."

79. Letter to Regius, January 1642, AT 3:491–92. Descartes did not always keep his own advice about the denial of substantial forms. When the authorities at Louvain wanted to condemn some Cartesian propositions, including Descartes's denial of substantial forms (see Chapter 10), they could not find the rejection of the doctrine in the *Principles* but were able to cite a passage from *Replies* VI, sec. 7.

[7]

Descartes and the Jesuits of La Flèche: The Eucharist

In his *Brief Lives of Contemporaries,* John Aubrey reports that Hobbes "was wont to say that had M^ieur^ Des Cartes (for whom he had great respect) kept himselfe to geometrie, he had been the best geometer in the world; but he could not pardon him for his writing in defence of transubstantiation, which he knew was absolutely against his opinion (conscience) and donne meerly to putt a compliment [on] (flatter) the Jesuites."[1] Thus, Aubrey maintains what became a popular interpretation of Descartes's views on the relationship between theology and philosophy and on Descartes's motives for entering into a theological discussion—specifically the one concerning the explanation of the sacrament of the Eucharist. One can find variations of this interpretation almost everywhere; it is frequent in pre-twentieth-century commentaries and still very common in Anglo-American scholarship.

In its usual form it goes as follows: Descartes did not want to bother with transubstantiation, but, after Arnauld's objections, he could not avoid it. The nineteenth-century commentator Francisque Bouillier, paraphrasing Descartes's late-seventeenth-century biographer, Adrien Baillet, states: "no doubt [Descartes] would truly have wanted to avoid meddling into this topic of transubstantiation, but, after Arnauld's objection, he was no longer allowed to remain in silence."[2] Descartes is said to have been dragged into a debate in which he did not want to participate and in which perhaps he

---

1. Aubrey 1898, 1:367.
2. Bouillier 1868, 1:450. It seems to be his opinion, though he puts it forward as Baillet's. He quotes Baillet, who writes: "Ce n'est pas que M. Descartes ne prit toutes les mesures possibles pour se dispenser de jamais remuer la matiere qui concerne la transsubstantiation au Sacrement de l'Eucharistie, parce qu'il la regardoit comme une question de pure Théologie, et comme un mystere que Dieu nous propose de croire sans nous obliger a l'examiner. Mais,

[140]

should not have participated. Baillet states: "It would have been desirable for Descartes to have recognized in good faith and steadfastly the moral impossibility in which all philosophers will always be, of demonstrating transubstantiation using the principles of physics, or that he would have had the strength to keep a perpetual silence on this point, not attempting to plumb the depths of so inexplicable a mystery."[3] In a recent article, Nicholas Jolley echoes: "instead of contenting himself with saying that the dogma [of transubstantiation] was a mystery that must simply be accepted on faith, Descartes attempted to explain it in terms of his own philosophy. Descartes's possibly misguided efforts were to be taken up by his overzealous disciples."[4]

As evidence for the general view, one may cite the alleged fact that Descartes issued different explanations of transubstantiation, suggesting that he was constructing his explanations as he went along; for instance, in recent work, Richard Watson counts three separate theories of transubstantiation: the first in Descartes's *Reply* to Arnauld, the second, "which turns out to be an explanation of transubstantiation completely different from the one he gives to Arnauld," in his letters to Denis Mesland, and the third when, not remaining "long satisfied that [the second theory] alone is adequate," Descartes, in a later letter to Claude Clerselier, "combines his first with his second theory."[5] Also adduced as evidence is the fact that churchmen who accepted Descartes's explanation got into trouble with Church authorities (as if they and Descartes were doing something wrong). Bouillier asserts: "A small time after this letter [on transubstantiation], Father Mesland was sent to the missions to tend to the savages, perhaps because of his overly ardent taste for the new philosophy. . . ."[6] And Watson tells us more forcefully that "the exchange of letters [between Descartes and Mesland] began in 1644 and was terminated abruptly in 1646 when, as extreme discipline for his commerce with Descartes, Mesland was banished to Canada."[7] He even asks, "Why was Mesland dealt with so severely?" and he answers, "Undoubtedly it was for the same reasons that led Descartes to

---

depuis que M. Arnauld luy en fait l'objection, comme au nom des Théologiens Scholastiques, il ne luy fut plus libre de demeurer dans son silence" (Baillet 1691, 2:520).

3. Baillet 1691, 2:521.
4. Nicholas Jolley, "The Reception of Descartes' Philosophy," in Cottingham 1992, 393–438.
5. Watson 1987, 161; Watson 1982, 135–36.
6. Bouillier 1868, 451. A marginal note on the manuscript copy of the letter to Mesland makes the same point: "Ce Pere fut relegué en Canada, ou il est mort, à cause de la trop grande relation qu'il auoit avec Mr Des Cartes." See AT 4:345.
7. Watson 1987, 156; 1982, 129. Watson adds that Mesland "died on the Canadian mission in 1672 without, as far as is known, inquiring further into transubstantiation."

drop his guard to make some tentative proposals about Cartesian theology himself. The issue of transubstantiation was crucial."[8]

Now, it is true that during the second half of the seventeenth century, Descartes and the Cartesians were very heavily criticized by various Scholastics, and especially by the Jesuits, for their explanations of the Eucharist. The issue seems to have been the focus of opposition to Cartesianism. It was alleged to be the cause of Descartes's works being placed on the *Index of Prohibited Books* in 1663; it was the issue to which a 1671 edict by the king of France referred; and it was specifically cited as grounds for the condemnation of Cartesianism by the University of Louvain in 1662. Among the disputed propositions at Louvain were those supporting the Cartesian rejection of substantial forms or real accidents.[9] The objection against these propositions was that, as a consequence, the accidents of bread and wine would not remain without subject in the Eucharist.[10] As early as 1665, one can find, as part of a general assessment of the doctrinal difficulties of Cartesianism by the Jesuits of the Collège de Clermont, that "the Cartesian hypothesis must be distasteful to . . . theology . . . because it seems to follow from the hypothesis that there can be no conversion of bread and wine in the Eucharist into the blood and body of Christ, nor can it be determined what is destroyed in that conversion, which favors heretics."[11] Moreover, in 1678 the Oratorians and the Jesuits decided to require their professors to teach *against Descartes* that "in each natural body there is a substantial form really distinct from matter" and that "there are real and absolute accidents inherent in their subjects, which can supernaturally be without any subjects."[12] At the same time at the Collège d'Angers, the Oratorian Fathers Fromentier, L'Amy, and Villecroze were removed from their teaching positions for having taught "the opinion of the Cartesians who state that there are no species or real accidents in the Eucharist, which is contrary to the theology of the holy fathers and to the doctrine of the church . . . and which was censured by the Sorbonne in 1624 as bold, erroneous, and approaching heresy."[13] Subsequently, the Jesuits formally condemned the following propositions: "There are no substantial forms of bodies in matter" and "There are no absolute accidents."[14] Many school books contained discussions of the above doctrines, as this example from Jean-Baptiste de la Grange's 1682 textbook shows: "the manner in which Descartes explains

---

8. Watson 1987, 156; 1982, 129.
9. *Principia* I, art. 51–51; *Responsiones* VI, sec. 7.
10. Ariew 1994a, 3; trans. in ACS 255.
11. As reported by Robert Boyle in a letter to Henry Oldenburg (Oldenburg 1966, 2:35).
12. See Ariew 1994a, 4; trans. in ACS 257.
13. Babin 1679, 39, 44.
14. Ariew 1994a, 6; trans. in ACS 260.

the Mystery of the Holy Eucharist is completely false, as we shall show in the present treatise on accidental forms. . . . If it is true, for example, that there are no substantial forms, as Descartes assumes, one cannot say that man would be rectified by an inherent grace—which is however what is set forth by the Council of Trent against the heretics."[15] With the publication in 1680 by the Père de Valois, writing under the pseudonym Louis de la Ville, of *Sentimens de Monsieur Descartes touchant l'essence et les proprietez du corps opposez à la doctrine de l'Eglise, et conforme aux erreurs de Calvin sur le sujet de l'Eucharistie*, discussions about the Eucharist became even more frequent but usually shifted from the Cartesians' denial of substantial forms and real accidents to the consequences of their principle that quantity or extension is corporeal subsistence.[16] The authorities at Louvain had previously found offensive the Cartesian principle that the extension of bodies constitutes their essential and natural attribute.[17] Oratorians and Jesuits were required to teach "that actual and external extension is not the essence of matter."[18] In 1691 the University of Paris condemned the proposition that "the matter of bodies is nothing other than their extension and one cannot be without the other."[19] The Jesuits formally echoed this sentiment with a prohibition of the proposition that "the essence of matter or of body consists in its actual and external extension."[20] Scholastic textbooks from the second half of the seventeenth century were filled with such discussions.[21]

It is also true that the Cartesians found Descartes's explanations in two of the Letters to Mesland so sensitive that they did not disseminate them widely. Clerselier did not publish these letters in his three-volume edition of Descartes's *Correspondance* (though he did circulate them in private).[22] Moreover, Descartes himself recognized that he was dealing with delicate matters. When writing to Mesland, he expressed the fear that since he was not a theologian by profession, things he might write could be less well taken from him than from another. Thus, he wrote about the Eucharist to Mesland "under the condition that if you communicated it to others, it would be without attributing its authorship to me, and even that you would

15. Grange 1682, 1:3; see also 1:109–35.
16. See Chapter 9.
17. Ariew 1994a, 3; trans. in ACS 255.
18. Ariew 1994a, 4; trans. in ACS 256.
19. Ariew 1994a, 4; trans. in ACS 257.
20. Ariew 1994a, 6; trans. in ACS 259.
21. For example, Duhamel 1692, 189–201. For more on this issue, see Chapters 8 and 9.
22. They were not actually published until the nineteenth century, first in Descartes 1811, and then, in a better edition, in Bouillier 1868, 453–59. In 1660, the Benedictine Dom Antoine Vinot advised Clerselier not to correspond with the Jesuit Jean Berthet: "you could not deliver a more deadly blow to the philosophy of M. Descartes . . . than to communicate your views on the Eucharist to those people the Jesuits" (quoted in McClaughlin 1979, 571). See also Nadler 1988, esp. 256.

communicate it to no one, if you judged that it is not completely in conformity with what has been determined by the Church."[23] Similarly, when Arnauld later asked Descartes for further explanations, Descartes answered that he would fear an accusation of rashness if he dared to come to any specific conclusion on the question, and that he would prefer to communicate such conjectures by word of mouth rather than in writing.[24] Such circumspection might have been needed; the sacrament of the Eucharist was a point of division between Catholics and Protestants at a time when religious wars had been waged throughout much of Europe.

But still, we should not read our history backward. Descartes could not have known that he was bound to fail. So I propose to read the letters to Mesland in the context of Descartes's correspondence with the Jesuits in general and in the broader context of discussions of the Eucharist before the 1640s. Descartes would then be seen as making genuine attempts to enlist the Jesuits into teaching his philosophy, attempts that follow the same practices as those of previous natural philosophers, attempts that will fail, of course, but that need not have failed. In this way, we can see that the common view of Descartes on the Eucharist is false on almost every count, and we can also reaffirm the interpretations of contemporary French commentators on Descartes and the Eucharist, from Henri Gouhier to Jean-Robert Armogathe.[25]

Cartesianism was not alone in being censured for holding doctrines inconsistent with various Church dogmas. Ironically, most of the difficulties with Cartesianism in the seventeenth century were previously difficulties with Aristotelianism in the thirteenth century. Among the propositions condemned by the bishop of Paris in 1277 were some that were seen as threatening to the Eucharist.[26] Moreover, in a notorious case in 1624, the Université de Paris and the Parlement prohibited the denial of substantial forms by some anti-Aristotelians on the grounds that holding an atomist philosophy would have been inconsistent with giving an intelligible explanation of transubstantiation.[27] The university prohibited several theses concerning matter and form, one in particular denying all substantial forms, except for the rational soul, along with prime matter; its official condemnation was that "this proposition is overly bold, erroneous, and close to

---

23. AT 4:165. In the next letter, Descartes suggests: "Vous ferez de ma lettre ce qu'il vous plaira, et pource qu'elle ne vaut pas la peine d'estre gardée, ie vous prie seulement de la rompre, sans prendre la peine de me la renvoyer," AT 4:216.

24. "Conjecturas autem meas viva voce malim exponere, quam scriptis," AT 5:194.

25. Gouhier 1972 and Armogathe 1977. Gaston Sortais (1929, 1937) can be seen as an early proponent of approximately the same view.

26. See propositions 196–99 (originally 138–40) in Mandonnet 1908, 175–91.

27. See Chapter 4. See also Garber 1988, 471–86; 1994; Blair 1993.

heresy."[28] There are many extant reports about the events of 1624, including some by Descartes's correspondents Mersenne and Jean-Baptiste Morin. These reports have little favorable to say about the anti-Aristotelians and their theses. Mersenne defends Aristotle and dismisses them as charlatans; all the reports evince concerns about the compatibility of anti-Aristotelian philosophy and Catholic theology.[29]

Moreover, the anti-Aristotelians were not being singled out in this respect. It was a common tactic at the start of the seventeenth century to claim that a particular philosophical view was not able to accommodate the Eucharist. For instance, Scipion Dupleix argues in his *Physique* that Thomists are inconsistent with their explanation of the Eucharist if they deny that matter can be without form;[30] that is, he argues against Thomas Aquinas's doctrine of prime matter as pure potency by analogy to what is required in the sacrament of the Eucharist:

> [Saint Thomas] truly agrees that God can make an accident subsist in nature outside its subject, in the same way that all genuine Christians believe that all the accidents of the bread are without the bread in the Holy Sacrament of the Eucharist, and the accidents of the wine are without the wine, even though it seems that there is a greater incompatibility in this than in having matter subsist without form, insofar as matter does not need any subject or support, being itself the subject and support of all other natural things, and accidents cannot naturally subsist without subjects.[31]

We can see the same line of argument in later textbooks. For example, René de Ceriziers argues, in *Le philosophe français,* that there can be no form without matter and no matter without form naturally, but he adds, "however, one must not deny that God can conserve matter without any form, since these are two beings that can be distinguished, that no more depend upon one another than accident on substance, the former being separated from the latter in the Eucharist."[32]

Such discussions were especially frequent in the commentaries on Aristotle's *Physics* concerning matter, form, place, time, and the void. We have just seen the Eucharist invoked by Dupleix and de Ceriziers in their discussions of matter, and by various authorities in Paris in their censure of the rejection of substantial form. We can also see Dupleix arguing, in his discussion of place, that supernaturally two bodies can be in the same place,

---

28. De Launoy 1656, 310–21. This prohibition was renewed in 1671 and became the basis for condemnations of Cartesianism; see Chapter 10.
29. Mersenne 1625, 100–101.
30. Dupleix 1990, 131–32.
31. Dupleix 1990, 131–32.
32. Ceriziers 1643, chap. 3, pp. 51–52.

and that, given the sacrament of the Eucharist, one body can be in two places.[33] This is a common discussion in early-seventeenth-century philosophy textbooks. Both of the questions—whether one body can occupy two places and whether two bodies can occupy one place—are answered affirmatively, given the Eucharist, by the Jesuits of the University of Coimbra and by Charles d'Abra de Raconis.[34] Another textbook writer, Eustachius a Sancto Paulo, holds a similar doctrine. After maintaining that two bodies can be in one place by divine virtue, he argues that there is no incompatibility involved in one body existing in several places. The example he gives for the latter proposition is that in the Sacred Eucharist the body of Christ is really and personally in several places.[35] Eustachius also evokes the Eucharist in his chapter on matter and quantity: "just as one and the same piece of matter can undergo various forms in succession, so one and the same quantity may endure in all these forms—and sometimes indeed there are changes in the very nature of the matter, as happens in the most revered sacrament of the Eucharist."[36]

Another early textbook writer, Théophraste Bouju, in his *Corps de toute la philosophie,* argues for the possibility of a void on the model of transubstantiation; he asserts the impossibility of internal place or space being void of all bodies, but he adds, "Except that God by his absolute power can give subsistence to quantity as he does, in the Holy Eucharist, to the species of bread and wine which remain after transubstantiation."[37] Even Pierre Gassendi, in 1624, accepting the seemingly innocuous doctrine that "the essence of quantity is nothing but its external extension,"[38] feels compelled to point out that his doctrine has negative consequences for the sacrament of the Eucharist and to take steps to reaffirm his orthodoxy: "To continue, let us now turn our attention to the famous difficulty concerning the

33. "Pour le regard de l'autre question, à sçavoir-non si un corps peut estre en divers lieux en mesme temps, je croy que naturellement cela ne se peut faire non plus que plusieurs corps ne se peuvvent trouver en mesme temps en un mesme lieu: mais que par la toute-puissance de Dieu l'un se peut aussi bien que l'autre: je dy que Dieu peut tout les deux: et par ainsi (puis qu'il l'a voulu et l'a dit) que le corps de son fils est en tous les sacremens de la sainct sacrée Eucharistie, et en chaque petite piece d'iceux" (Ceriziers 1643, 261–62).

34. Conimbricenses 1594, lib. 4, cap. 5, quaest 4, art. 2, "Certum esse posse duo corpora virtute divina eodem loco simul existere"; quaest. 5, "Utrum idem corpus simul in duobus loci divina virtute esse queat," art 1, "Solutio quaestionis," esp. pp. 42, 44 (though their account is a strange hybrid of Thomism and Scotism). Raconis 1651, pars 3, Physica, tract. 2, "De loco, ad quartum librum physicorum," quaest. 1, "An plura loca idem numero corpus capere possint, seu an idem numero corpus possit esse in pluribus locis"; quaest 2, "An duo vel plura corpora possint esse in eodem loco per penetratione," esp. pp. 207, 216.

35. Eustachius a Sancto Paulo 1629, Physica, pars 1, tract. 3, disp. 2, quaest. 3, "An duo corpora in eodem loco, et idem corpus in duobus locis esse possit," p. 59.

36. Eustachius 1629, Physica, pars 1, tract. 1, disp. 2, quaest. 4, "Quaenam sint praecipuae proprietatis materiae," p. 16.

37. Bouju, 1614, 469.

38. Gassendi 1624, II, exer. 3, art 10.

essence of quantity. Our philosophers explain it so well that nothing could be more obscure, though nothing would seem to be more obvious than quantity. However, I must confess that the mystery of the Eucharist, as our faith conceives it, may cause some difficulty in this matter."[39]

The inescapable conclusion is that, at least during the first half of the seventeenth century, it was the common practice of Catholic philosophers when they were theorizing about natural philosophy to discuss the compatibility of their physical theories with such mysteries of the Catholic faith as the sacrament of the Eucharist. The question is what Descartes's correspondence might reveal about the extent of his awareness of these practices. An initial response might emphasize that Descartes read the philosophy textbooks of at least two of the above scholastic authors—Eustachius a Sancto Paulo and Charles d'Abra de Raconis—in November 1640, just before his first pronouncements on the Eucharist in *Replies* IV to Arnauld, circa March 1641.[40] He surely knew some of the others as well: he was taught from the textbooks of the Conimbricenses at La Flèche.[41] Thus, a more careful examination is called for.

Descartes's correspondence with the Jesuits can be considered as three separate series of letters, each spanning a couple of years. First are four letters to some Jesuits at La Flèche in 1637 and 1638—possibly Noël, Fournier, and Vatier—requesting comments about his newly published *Discourse on Method*.[42] Second is the series of letters written during 1640 to 1642, dealing with Descartes's "war with the Jesuits," that is, the Bourdin affair, and culminating with the *Letter to Dinet,* published with the second edition of the *Meditations*.[43] And third is the set of letters from 1644 to 1646, predominantly involving Mesland, but also including others such as Charlet, Noël, Grandamy, and the now friendly Bourdin.[44] For most of these letters, Clerselier does not provide the name of the correspondent or the date of the letter; he simply identifies them as "A un reuerend Pere Iesuite." Descartes himself generally treats the Jesuits as if they were a collective whole, that is, as if it did not really matter with whom he was corresponding. In the *Seventh Objections,* he refers to the Jesuits as "a society

---

39. Gassendi 1624, II, exer. 3, art. 10; also art 11, "Species Eucharisticas non item fore Fides nos Orthodoxa docet."
40. See letter to Mersenne, 11 November 1640, AT 3:232–34, for the references to Eustachius and Abra de Raconis, and letter to Mersenne, 4 March 1641, AT 3:328, for the first mention of the replies to Arnauld.
41. See AT 3:185; see also Chapter 1.
42. AT 1:382–84, 454–56, 456–58, 558–65.
43. AT 3:97–100, 105–18, 168–74, 221–28, 269–77, 464–68, 575–77, 594–602, 7:563–603.
44. AT 3:378–82, 4:110–20, 121–23, 139–41, 142–43, 143–44, 156–58, 158–60, 160–61, 161–70, 172–79, 215–17, 344–48, 584–86, 587–88.

which is very famous for its learning and piety, and whose members are all in such close union with each other that it is rare that anything is done by one of them which is not approved by all";[45] and he says to Huygens: "since I understand the communication and union that exists among those of that order, the testimony of one of them alone is enough to allow me to hope that I will have them all on my side."[46]

The first four letters of 1637–38 are extremely interesting. Writing to one of his old teachers—something he has not done for at least twenty years—Descartes sends him the *Discourse on Method* as "a fruit that belongs to him, and whose first seeds were sown by him in his mind, as he owes to those of his Order the little knowledge he has of letters."[47] Descartes asks the Jesuits to examine the work. In the second and third letters, Descartes again asks for "censures" and asks the Jesuits to continue to teach him: "For there is no one, it seems to me, who has more interest in examining this book than the members of your Society. . . . I do not know how they can henceforth continue to teach these subjects, as they do every year in most of your colleges, unless they disprove what I have written or unless they follow it."[48] It is in this context that Descartes communicates his awareness that the principal reason the Jesuits reject all kinds of novelties in philosophy is the fear that these might also cause some change in theology.[49] Some four years before the *Replies* to Arnauld, Descartes boldly asserts in the letter to his Jesuit correspondent that his naturalistic explanations are consistent with the mysteries of the Catholic faith, that is, that his physics is compatible with Catholic revealed theology—including the mystery of the Eucharist—*and that he will make this clear in the future.* Even more to the point, in the fourth letter, Descartes, buoyed up by his correspondent's favorable reception to his *Essays*, tells him that his opinions were not conceived lightly and that they are worth the bother of examining. He then adds: "I say to you also that I do not fear that anything against the faith would be found in [my physics and metaphysics]; for on the contrary, I dare boast that faith has never been so strongly supported by human reasons as it may be if one follows my principles; transubstantiation, in partic-

---

45. AT 7:452. See also *Lettre au P. Dinet*, AT 7:64.
46. AT 2:50. See also Ariew 1995.
47. AT 1:383.
48. AT 1:455.
49. AT 1:455–56. Later, in *Lettre au P. Dinet*, Descartes repeats this claim: "Atque omnino profiteor nihil ad religionem pertinere, quod non aeque ac etiam magis facile explicetur per mea principia, quam per ea quae vulgo recepta sunt (j'avance hardiment que notre religion ne nous enseigne rien qui ne se puisse expliquer aussi facilement, ou même avec plus de facilité, suivant mes principes, que suivant ceux qui sont communément reçus)." He then refers to his striking example, at the end of *Responsiones* IV, of his ability to explain matters relating to religion better than the ordinary philosophy—and adds that he would be ready to do the same for any other topic, if needs be; AT 7:581.

ular, which the Calvinists take as impossible to explain by ordinary philosophy, is very easily explained by mine."[50]

Descartes is not only implying but declaring openly that he has an explanation of the Eucharist based on his principles of natural philosophy. This should not really surprise us; for, as early as 25 November 1630, Descartes, writing to Mersenne about his physics, specifically about his theory of colors, states: "I think I will send you this discourse on light as soon as it is done, and before sending you the rest of the *Dioptrics;* for, wanting to describe colors in it in my way, and consequently, being obliged to explain in it how the whiteness of the bread remains in the Holy Sacrament, I would feel more comfortable if it were examined by my friends before it is seen by the whole world."[51] Thus, it is clear that Descartes understood very well the practices of contemporary Catholic philosophers when they were theorizing about natural philosophy; as early as 1630, deliberating about color, he understood that he would be required to discuss the compatibility of his theory and the mystery of the Eucharist. It is probable that Descartes had worked out his explanation of the Eucharist as early as 1630, and that in 1637–38 he was sincere in his belief that his principles of physics were consistent with the mysteries of religion. His declaration in 1637 that he would make this clear on one or more occasions was unlikely to be an empty one. And, in 1638, he must have been ready to substantiate his boast that faith, particularly the mystery of transubstantiation, had never been so strongly supported by human reasons as it could be by his principles.

When Descartes first announced to Mersenne that he was sending him the *Replies* to Arnauld's objection concerning the Eucharist, Descartes seemed very confident of his position. "You will see that my philosophy agrees so much with what is determined by the Councils about the Holy Sacrament," he declared to Mersenne, "that I maintain that it is impossible to give a satisfactory explanation of it by means of the traditional philosophy. Indeed I think that the latter would have been rejected as repugnant to faith had mine been known first.... I am confident that I can show that there is no opinion in their philosophy that accords as well with faith as mine."[52] It is true that, in the first edition of the *Meditations* in 1641, Descartes refrained from publishing the last few paragraphs of his *Reply* to Arnauld; as he says, he censored himself at Mersenne's urging, so that he would not have any difficulty in getting the approbation of the Sorbonne.[53] He explains to Huygens, "Father Mersenne has pruned 2 or 3 pages from

50. AT 1:564.
51. Letter to Mersenne, 25 November 1630, AT 1:179.
52. AT 3:349–50.
53. "J'approve fort que vous auez retranché ce que i'auois mis à la fin de ma Réponse à M. Arnauld, principalement si cela peut nous aider à obtenir une approbation" (AT 3:416).

the end of my replies to the *Fourth Objections,* concerning the Eucharist, because he feared that the Doctors would be offended in that I proved there that their opinion concerning that point did not agree as well as mine with the Scriptures and the Councils."[54] In any case, Descartes quickly restored the pages in the second Amsterdam edition in 1642, when there was no longer any need for the approbation of those Doctors.[55]

Arnauld's objection is that transubstantiation requires that the accidents of the bread remain after the substance of the bread is taken away. According to Arnauld, this would not be possible in Cartesian philosophy, since for Descartes there are no real accidents, but only modes of substance that are unintelligible apart from the substance in which they inhere.[56] Descartes accepts this characterization of his position, except that he denies ever having rejected real accidents and affirms that God can bring about things that we are incapable of understanding.[57] But he then goes on to sketch an account of how objects affect the senses by means of their surface or the surrounding air or other bodies, opposing the scholastic theory of the transmission of intentional species.[58] He hypothesizes that if the substance of the bread is changed into the substance of something else but still occupies the boundaries occupied by the previous substance, the new substance will affect our senses in the same way the old one did.[59] Descartes even quotes from the Council of Trent and goes on to prefer his hypothesis to the teaching of the theologians (the section excised by Mersenne).[60]

Descartes's response is obviously limited to the problem at hand: to explain, without using real accidents, how the bread after transubstantiation might still look like bread to us. Even a superficial look at Thomas's *Summa Theologiae,* part 3, questions 73–78 (and 79–83), would indicate that there are many issues not discussed by Descartes and, in particular, that he has said nothing about the real presence of Christ in the consecrated bread. This is, in fact, what Descartes indicates to Mesland: "As for the extension of Jesus Christ in that Sacrament, I gave no explanation of it, because I was not obliged to, and I keep far away, as far as possible, from questions of the-

---

54. AT 3:772.
55. See AT 3:785, in which Descartes tells Huygens that the second edition of the *Meditations* is "plus correcte que celle de Paris, et mesme un peu plus ample, principalement en la fin de ma response aux quatriesmes obiections, ou ie me suis emancipé d'escrire que l'opinion commune de nos Theologiens touchant l'Eucharistie n'est pas aussi orthodoxe que la mienne, ce que le pere Mercenne auvoit retranché pour ne pas deplaire à nos Docteurs." See also Armogathe 1994.
56. AT 7:217–18. See also Stephen Menn, "The Greatest Stumbling Block: Descartes' Denial of Real Qualities," in Ariew and Grene 1995.
57. AT 7:248–49.
58. AT 7:249–51.
59. AT 7:251.
60. AT 7:251–56.

ology, especially as the Council of Trent has said that he is present, *ea existendi ratione quam verbis exprimere vix possumus* [in this manner of existing which we can barely express by means of words]. I quoted that phrase, toward the end of my Reply to the Fourth Objections, precisely to excuse myself from giving an explanation."[61] He adds, however: "But I venture to say that if people were a little more used to my way of philosophizing, they could be shown a way of explaining this mystery which would stop the mouths of the enemies of our religion so that they could say nothing against it."[62]

Descartes provides his explanation in a subsequent letter. First he reaffirms and clarifies the hypothesis of the *Fourth Replies*,[63] then he sketches an account of individuation for bodies and the human body: "The numerical identity of the surface does not depend on the identity of the bodies between which it exists, but only on the identity or similarity of the dimensions. Similarly we can say that the Loire is the same river as it was ten years ago, although it is no longer the same water, and perhaps there is no longer even a single part of the earth which then surrounded that water."[64] In the case of a human body, it remains the same through changes of matter, on account of its union with a soul: "they are *eadem numero* [numerically the same], only because they are informed by the same soul."[65] Thus, humans naturally transubstantiate other matter by incorporating it and making it part of their bodies, bodies that are informed by a soul.[66] In a similar fashion, Descartes accounts for the miracle of transubstantiation by having the soul of Christ supernaturally inform the matter of the host upon consecration.[67]

There is no doubt that both aspects of Descartes's explanations of the Eucharist are in conflict with Thomist explanations. What is not often noted, however, is that seventeenth-century Catholic explanations of the Eucharist were as much in conflict with Thomist explanations as were Descartes's. As we have seen, the explanation of the sacrament of the Eucharist refers not only to specific metaphysical and physical theories about

---

61. 2 May 1644, AT 4:119–20. See also AT 4:374–75.
62. AT 4:120.
63. 9 February 1645, AT 4:163–64.
64. AT 4:164–65.
65. AT 4:167.
66. AT 4:167–68.
67. AT 4:168–69. See also letter to Mesland, 1645 or 1646, AT 4:345–48, and letter to Clerselier, 2 March 1646, AT 4:371–73. Clearly, an explanation of how any object affects our senses is still an explanation in the realm of philosophy, for Descartes; so is an explanation of the principle of individuation for bodies (and "natural transubstantiation"). However, when Descartes explains transubstantiation as a supernatural phenomenon, he is entering into the realm of theology.

substance, substantial forms, accidents or modes, and extension and quantity, but also to place and to the principle of individuation. In school texts during the seventeenth century, there were (at least) two major alterations to the metaphysics underlying transubstantiation. The first—which we have already noted—is that Schoolmen accepted (against Thomas) that two bodies could be in one place and one body could be in many places at the same time, thus changing the interpretation of how Christ could be in the sacrament. This issue was a point of contention between Thomists and Scotists in their discussions of the Eucharist, Thomas formally rejecting the notion that one body can be in two places[68] and Scotus officially insisting that two bodies can be in one place.[69] Ironically, Descartes ranks himself with Thomists on this issue. Though he does not state it explicitly, it is clear that his reasoning in *Principles* II, art. 22, on the impossibility of multiple worlds, requires the impossibility of two bodies being in one place at the same time.[70]

The second major alteration in seventeenth-century Scholasticism is that the principle of individuation changed from matter to form, thus modifying the grounds on which one says that Christ is in the sacrament.[71] For example, Scipion Dupleix's position is the Scotist opinion that "in order to establish the individual essence of Socrates, Alexander, Scipion, and other singular persons, we must necessarily add for each one of them an individual and singular essential difference which is so proper and so peculiar to each of them for themselves, that it makes each of them differ essentially from all the others."[72] And there are similar doctrines in the Metaphysica of Eustachius a Sancto Paulo and that of Abra de Raconis.[73] The Scotist posi-

---

68. Aquinas 1964–76, pars 3, quaest. 75, "Quia impossibile est quod unus motus ejusdem corporis localiter moti terminetur simul diversa loca"; see also Aquinas 1918–30, IV, chap. 63. For Thomas's denial that one body can be in two places, see Aquinas 1964–76, pars 3a, quaest. 83, 84; *Quodlibeta* I, art. 2; *Physica* IV, lect. 9 and *Metaphysica* III, lect. 7.

69. Scotus 1968, *Quaestiones quodlibetales*, quaest. 10, art. 2.

70. This was pointed out in the seventeenth century

Ce que je trouve de plaisant, c'est que Descartes enseigne hardiment des conclusions très dangereuses, qu'il tire de deux principes qui ne sont point prouvez. Le premier principe qu'il suppose, est que par tout il y a de l'espace, il y a aussi de la matière.... Le second principe qu'il doit supposer nécessairement, pour conclure que plusieurs mondes sont impossibles, et dont neantmoins il ne parle point; c'est que deux corps ne peuvent pas, absolument parlant, estre dans un mesme lieu, et que la matière ne peut pas estre dans une autre matière.... De sorte qu'il faut remarquer que non seulement la conclusion de Descartes, que plusieurs mondes sont impossibles est fausse et dangereuse; mais aussi qu'elle est tirée d'un principe dangereux, qui est que deux corps ne sauraient estre, absolument parlant, dans le mesme espace. (Grange 1682, 7–9)

71. See Chapter 4.
72. Dupleix 1992, 235.
73. Raconis 1651, pars 4, Metaphysica, tract. 4, sec. 2, 4, brevis appendix, "De unitate singulari et numerica, seu principio individuationis," pp. 76–78; Eustachius a Sancto Paulo

tion seems to be the majority position in the seventeenth century. It entails that form, not matter, is the principle of individuation. Thus Descartes's position in this second respect looks very much like the majority position.

In spite of the pronouncements of the Council of Trent (or perhaps because of them) there was a fair degree of latitude in the first half of the seventeenth century with respect to the metaphysical issues that provided the foundations for explanations of the Eucharist.[74] Obviously, there were clear indications that certain views were prohibited—consubstantiation, for example—but there did seem to be enough room for Descartes to be hopeful that his views would be accepted.[75]

Of course, all of this simply belies the common interpretation. The fact is that Descartes's account of transubstantiation was not done merely to flatter the Jesuits, as Hobbes believed; Descartes considered the relationship between his physics and Catholic theology years before he corresponded with the Jesuits. Nor did he want to avoid meddling in the topic, as Baillet and Bouillier stated. As Descartes himself tells us, it was the Calvinists who took transubstantiation to be impossible to explain by ordinary philosophy. Thus, Descartes could not have accepted Jolley's suggestion that he should have contented himself with saying that the dogma must simply be accepted on faith. Nor could he have accepted Baillet's similar pronouncement that philosophers will always find it impossible to demonstrate transubstantiation using the principles of physics. One needs to understand Hobbes's statements as more revealing of his Protestant views than of Descartes's genuine motives; similarly, Baillet's comment tells us more about Jansenist (and anti-Jesuit) politics at the end of the seventeenth century than about Descartes.[76] Further, since Descartes probably worked out an account of transubstantiation based on his principles of physics well before his *Replies* to Arnauld, it seems doubtful that he would be making up

---

1629, pars 4, Metaphysica, tractatus de proprietatibus entis, disp. 2, de simplicibus proprietatibus Entis, quaest. 4, "Quodnam sit principium unitatis numerica, seu individuationis," pp. 38–39. Another doctrine in the same general direction is that of Franco Burgersdijk (1657, I, chap. 12, "De Unitate Numerica et formali, deque principio individuationis," pp. 66–75). Burgerdijk is obviously not a French Catholic philosopher, but he does represent a natural extension of the Scotist view, which almost merges with the Cartesian doctrine. He rejects both Scotus's and Thomas's opinions (with Thomas's as the worse one). His own doctrine is that composite substances have both material and formal principles of individuation (pp. 71–72). With humans, the individuality lies in the rational soul, which is an immaterial form. And, of course, humans can also be differentiated by their accidents (pp. 74–75).

74. See Lewis 1995.

75. See Nadler 1988 on Arnauld's attempts to defend Descartes's philosophy (including the explanations of the Eucharist) against various attacks in the latter half of the seventeenth century.

76. See Gregor Sebba, "Adrien Baillet and the Genesis of His *Vie de M. Des-cartes*," in Lennon, Nicholas, and Davis 1982, 9–59.

theories as he went along. Thus Watson's interpretive theses are unlikely as well. Finally, it is improbable that Mesland was being disciplined by the Jesuits for his commerce with Descartes. We cannot be sure, of course, but we can surmise that being sent to the missions was not a punishment but a reward for a Jesuit in the seventeenth century. Camille de Rochemonteix states (though without giving any documentation) that Mesland had requested the assignment himself.[77] Moreover, the length of Mesland's stay in the New World (at least twenty-six years, more than twenty-two years after Descartes's death) is evidence against the thought that Mesland was being punished—or if this was punishment, it was very severe punishment indeed![78] The only plausible account for the whole affair is to think that Descartes was attempting to enlist the Jesuits as teachers of his philosophy and making a genuine attempt to establish his philosophy as a new Catholic philosophy to replace the Aristotelian—with all that this entails. Together with the *Principles of Philosophy*, Descartes's correspondence with Mesland constitutes a first step toward a Cartesian Scholasticism.[79]

77. De Rochemonteix 1889, 4:78.
78. For details about Mesland's stay in the New World—in Martinique and Santa Fe (now Bogota) of Nouvelle-Grenade (now Columbia), not Canada—see AT 4:669 and Armour 1988; both provide basic data to counter the standard misinformation about Mesland (especially that of the *Bibliothèque de la Compagnie de Jésus*, which confuses Denis Mesland with Pierre Mesland, who also taught at La Flèche and who died before the exchange of letters with Descartes).
79. In this respect, see, in particular, Descartes's letter "A un reverend Pere Iesuite," AT 4:121–22, in which he repeats that at La Flèche, in his youth "i'ay receu les premieres semences de tout ce que i'ay iamais appris, dequoy i'ay toute l'obligation à vostre Companie," and talks about Mesland's paraphrase of the *Meditations*, which will be "efficace pour authoriser mes Meditations." Cf. also the letter of 14 December 1646, AT 4:584–86.

[8]

Condemnations of Cartesianism: The Extension and Unity of the Universe

Descartes made some converts to his new philosophy with the publication of the *Principles,* the systematic exposition of his thought set out in scholastic style, but on the whole he did not succeed in getting the work adopted in the curriculum of the schools. Here and there, one can find Cartesian principles taught, as with the ill-fated Oratorians at Angers in the 1670s and Edmond Pourchot at Paris in the 1690s. One can also find Cartesian propositions included in some disputations, but the discussion is mostly negative. For most of the seventeenth century, the official response to Descartes's philosophy was unfavorable. At various times, Descartes waged fierce battles with his opponents. In the 1640s, he thought himself at war with the Jesuits.[1] And there were troubles and official condemnations by Protestants at Utrecht around 1642 and at Leyden in 1647.[2] The battles continued and intensified after Descartes's death in 1650. There were condemnations by Catholics at Louvain in 1662,[3] culminating with Descartes's works being put on the *Index of Prohibited Books* by the censors of Rome in 1663.[4] The fighting raged in the second half of the seventeenth century: the Jesuits held more anti-Cartesian disputations at Collège de Clermont in 1665, some clearly intended to make Descartes look ridiculous.[5] It intensified with numerous attacks in print.[6] The Cartesians counterattacked with satires[7] and learned

---

1. See Chapter 1.
2. Verbeek 1992.
3. Argentré 1736, pt. 2, pp. 303–4.
4. Bouillier 1868, 1:446–47.
5. Prou 1665.
6. See Vincent 1677; Ville [Louis le Valois] 1680; Grange 1682; and Huet 1689.
7. See the "arret burlesque," Boileau 1747, 3:150–53; Murr 1992, 231–40.

essays.⁸ The anti-Cartesians also responded with their own satires.⁹ Ultimately, the dispute spilled into the official political arena, the domains of the king, of the universities, and of the teaching orders. The king issued an edict in 1671;¹⁰ the Faculty of Arts at Paris tried to condemn Cartesianism in 1671 and succeeded in 1691;¹¹ there were skirmishes at Angers and Caen during 1675–78;¹² the Jesuits, in a Congress with the Oratorians, ultimately prohibited the teaching of Cartesianism in 1678,¹³ and formally condemned it in 1706.¹⁴

The official condemnations of Cartesianism of the late seventeenth century were unusually frequent and ferocious. Only the condemnations of Aristotelianism in the thirteenth century seem to have been as frequent and as wide sweeping. The reasons for the prohibitions of Cartesianism were even more diverse than those given against Peripatetic philosophy, however. Cartesianism was censured not only for doctrinal reasons, but also on pragmatic and pedagogical grounds. Reflecting the pedagogical judgment of the authorities of Utrecht, it was often asserted that being taught Cartesian philosophy would leave one unprepared for the higher faculties of theology, law, and medicine.¹⁵ And the Jesuits, echoing Pierre Bourdin's preoccupation with hyperbolic doubt, usually gave pragmatic reasons for dispensing with Cartesianism. This view can be captured nicely by the following comment by René Rapin: "In truth, Descartes teaches one to doubt too much, and that is not a good model for minds who are naturally credulous; but, in the end, he is more original than the others."¹⁶ A general assessment of the doctrinal difficulties of Cartesianism can be found in a summary of a disputation by the Jesuits of Clermont College during 1665:

> To say no more, the Cartesian hypothesis must be distasteful to mathematics, philosophy, and theology. To philosophy because it overthrows all its princi-

---

8. [Antoine Arnauld?], *Plusieurs raisons pour empecher la censure ou la condemnation de la philosophie de Descartes*, in Boileau 1747, 3:117–41 (reprinted in Cousin 1866, 3:303–17). See also Bayle 1684.
9. Daniel 1690; A. [Pierre Daniel Huet] 1692; Daniel 1693.
10. Bouillier 1868, 1:469.
11. Argentré 1736, pt. 1, p. 149.
12. For an account of the events at Angers, see Babin 1679.
13. *Concordat entre les Jesuites et les Peres de l'Oratoire, Actes de la Sixiéme Assemblée, September 1678*, in Bayle 1684, 11–12.
14. Rochemonteix 1889, 4:89–93. The full text of the documents in notes 10–14 is given in Ariew 1994a, 1–6; trans in ACS 254–60.
15. See Verbeek 1992; Ariew forthcoming.
16. Rapin 1725, 366. Doubt is often the target of criticism. See Babin 1679, 41: "Dire qu'il faut douter de toutes choses, c'est un principe qui tend à l'athéisme . . . ou du moins l'hérésie des manichéens"; cf. also the condemnations of 1691 and 1705, propositions 1–4 (Ariew 1994a, 5; trans in ACS 258). For the textbook critiques, see Vincent 1677, 3–12, and Duhamel 1692, 1–8.

ples and ideas which commonsense has accepted for centuries; to mathematics, because it is applied to the explanation of natural things, which are of another kind, not without great disturbance of order; to theology, because it seems to follow from the hypothesis that (i) too much is attributed to the fortuitous concourse of corpuscles, which favors the atheist; (ii) there is no necessity to allow a substantial form in man, which favors the impious and dissolute; (iii) there can be no conversion of bread and wine in the Eucharist into the blood and body of Christ, nor can it be determined what is destroyed in that conversion, which favors heretics.[17]

This summary is broken down into three main categories: the first, a complaint already issued at Utrecht, is the rejection of any novel philosophy. Descartes had previously attempted to defend himself against that charge by arguing (unsuccessfully, it seems) that his philosophy was not novel but the oldest of all philosophies, since he accepted only those principles generally admitted by all philosophers.[18] The second refers to the scholastic doctrine of the classification of the sciences. The claim is that mathematics should be subordinated to physics and not vice versa, as with Descartes. The third is itself divided into three parts, all concerning the relations between philosophy and theology. Cartesian philosophy is unfairly linked with atomism and the standard complaint against atomism is issued against it.[19] The disputants also object that, for Descartes, man's substantial form is not necessary, something Descartes himself complained about with respect to Regius's exposition of his philosophy.[20] At last, we come to the issue of the Eucharist, which seems to have been the focus of opposition to Cartesianism from at least 1671 on. It was the issue to which the king's edict referred, it was alleged to be the cause of Descartes's works being placed on the *Index;* and it was specifically cited as a ground for condemnation at Louvain, along with a few other intriguing difficulties.

---

17. A report of the disputation can be found in a letter from Oldenburg to Boyle of 4 July 1665, Oldenburg 1966, 2:430–35; the summary is on p. 435.
18. AT 7:580–81, 596. See Ariew 1994a. The accusation, however, continued to be repeated; see for example, Grange 1682, 1:1–2:

> Il ne faut qu'entendre Descartes expliquer les plus grands mystères de la Foy d'une manière toute nouvelle; et s'assurer que tous les Théologiens Catholiques se sont trompez jusqu'à présent, pour se persuader que si sa doctrine n'est pas erronée, du moins elle est dangereuse, et que les professeurs de philosophie ont tous les torts du monde de l'enseigner aux jeunes gens, a qui il est bon de ne point inspirer l'amour de la nouveauté, non plus que du mépris pour l'ancienne doctrine.

19. The same accusation can be found in Goudin 1864 [1668], vol. 2, p. 16, art. 3, "Des principes des choses suivant Leucippe, Democrite et Descartes"; and art, 4, "Exposé de la doctrine de Descartes sur les principes. 1. Opinion de Descartes sur les principes des choses 2. Les molecules de Descartes ne peuvent être les principes des choses."
20. Letter to Regius, mid-December 1641, AT 3:460–62; *Lettre au P. Dinet*, AT 7:585–86.

The 1662 condemnation at Louvain (which, according to Victor Cousin, was instigated by Jesuits) listed five difficulties with Cartesian doctrine: (1) Descartes's definition of substance, (2) his rejection of substantial forms or real accidents, (3) his doctrine that extension is the essential attribute of matter, (4) his espousal of the indefinite extension of the world, and (5) his rejection of the possibility of a plurality of worlds.[21] These five were heard again and again throughout the seventeenth century.

The definition of substance. The authorities at Louvain specifically referred to *Principles* I, article 51: "By substance we can understand nothing other than a thing which exists in such a way as to depend on no other thing for its existence. And there is only one substance which can be understood to depend on no other thing whatsoever, namely God. In the case of all other substances, we perceive that they can exist only with the help of God's concurrence"; and article 52: "But as for corporeal substance and mind (or created thinking substance), these can be understood to fall under this common concept: things that need only the concurrence of God in order to exist."[22] Their objection to these principles was that consequently there would be no substantial forms except for the rational soul—no substantial forms in animals and plants. The issue was picked up in the textbooks. One can find numerous discussions of the Cartesian definition of matter and body and repeated criticisms of the consequence that animals are machines lacking sensation and knowledge; these spanned such diverse thinkers as the Scotist Claudius Frassen and the Oratorian Jean-Baptiste de la Grange, among others.[23] Ultimately, the Jesuits condemned the proposition that "animals are mere automata deprived of all knowledge and sensation."[24]

The rejection of substantial forms or real accidents. Here the reference was to *Replies* VI, section 7, where Descartes stated:

> It is completely contradictory that there should be real accidents, since whatever is real can exist separately from any other subject; yet anything that can exist separately in this way is a substance, not an accident. The claim that real accidents cannot be separated from their subjects "naturally," but only by the power of God, is irrelevant. For to occur "naturally" is nothing other than to occur through the ordinary power of God, which in no way differs from his extraordinary power—the effect on the real world is exactly the same. Hence if

---

21. Ariew 1994a, 3; trans. in ACS 254.
22. AT 8:24–25; CSM 1:210.
23. See, for example, Frassen 1668, 30: "Rejicitur sententia Cartesii de materiae et corporis definitione"; "Negat Cartesius dari animam sensitivam atque cognoscitivam in brutis; et asserit esse meras machinas, quae ex sola organorum dispositione, et artificiosa partium structura instar horologii moventur," p. 646. Cf. also Grange 1682, 1:13: "Les bestes n'on point de raisonnement."
24. Ariew 1994a, 6; trans. in ACS 259.

everything which can naturally exist without a subject is a substance, anything that can exist without a substance even through the power of God, however extraordinary, should also be termed a substance.[25]

The objection was that, as a consequence, the accidents of bread and wine would not remain without subject in the Eucharist. This was surely the most frequently repeated criticism of Cartesianism. Oratorians and Jesuits required their professors to teach that "in each natural body there is a substantial form really distinct from matter," and "there are real and absolute accidents inherent in their subjects, which can supernaturally be without any subjects."[26] At Angers, the Oratorians Fromentier, L'Amy, and Villecroze were removed from their teaching positions for having taught the Cartesian doctrine that there are no species or real accidents in the Eucharist.[27] And the Jesuits condemned the propositions "There are no substantial forms of bodies in matter" and "There are no absolute accidents."[28] Most textbooks contained discussions of the doctrine.[29] But the discussions about the Eucharist shifted from the Cartesians' denial of substantial forms and real accidents to the consequences of their principle that quantity or extension is corporeal subsistence—that is, to objection 3.

Extension as an essential attribute of matter. The authorities at Louvain found offensive the Cartesian principle that the extension of bodies constitutes its essential and natural attribute. As with the next two principles condemned, they did no more than refer to the Cartesian text. For this tenet, they cited *Principles* I, article 53, and the *Meditations*,[30] though what they attributed to Descartes was not an exact quotation. Oratorians and Jesuits required one to teach "that actual and external extension is not the essence of matter";[31] in 1691 the University of Paris condemned the proposition "The matter of bodies is nothing other than their extension and one cannot be without the other";[32] and the Jesuits echoed with a prohibition of

---

25. AT 7:434–35; CSM 2: 293.
26. Ariew 1994a, 4; trans. in ACS 256–57.
27. Babin 1679, 39, 44. See also Chapter 7.
28. Ariew 1994a, 6; trans. in ACS 260.
29. For example, Grange 1682, 1:3; also 1:109–35. See also Chapter 7.
30. Descartes, *Replies* VI, AT 7:442 (CSM 3: 298): "I did not attribute to gravity the extension which constitutes the nature of a body"; *Principles* I, art. 53 (AT 8:25; CSM 1: 210): "each substance has one principal property which constitutes its nature and essence, and to which all its other properties are referred. Thus extension in length, breadth and depth constitutes the nature of corporeal substance."
31. Ariew 1994a, 4; trans. in ACS 256.
32. Ariew 1994a, 5; trans. in ACS 257.

"The essence of matter or of body consists in its actual and external extension."[33] The textbooks were filled with such arguments.[34]

The indefinite extension of the world. Here Louvain referred to *Principles* II, article 21: "What is more, we recognize that this world, that is, the whole universe of corporeal substance, has no limits to its extension."[35] Similarly, the Oratorian Fathers of Angers are said to have wrongly taught that "the world is infinite in its extension, a principle which is no less dangerous than the first [concerning the Eucharist],"[36] and the Jesuits condemned the proposition "In itself, the extension of the world is indefinite."[37] The issue was given full play in the textbook tradition.[38]

The plurality of worlds. The authorities at Louvain referred to *Principles* II, article 22: "It can also easily be gathered from this that ... if there were an infinite number of worlds, the matter of which they were composed would have to be identical; hence, there cannot in fact be a plurality of worlds, but only one."[39] Oratorians and Jesuits affirmed that one must teach "that there is no repugnance in God's creating several worlds at the same time,"[40] and the Jesuits condemned the proposition "There can be only one world."[41] Typically, the argument found in the textbooks was that Descartes was infringing on God's omnipotence: "For who would believe that Descartes teaches only the truth and what is known clearly by natural light, when he tells us in part 2 of his *Principles,* article 22, that several worlds are impossible? Can anything more novel and more shocking to reason be uttered? Ever since people have attempted to reason about God's works, possibly there has not been one who has dared to teach this doctrine, or even who has been of that opinion. In fact, there is nothing that seems more clear and natural to us than to assert that God, having produced this world, can still produce another."[42]

---

33. Ariew 1994a, 6; trans. in ACS 259.
34. For example, Duhamel 1692, 189–201. For more on this issue, see Chapters 9 and 11. The classic discussion of this problem is Armogathe 1977.
35. AT 8:52; CSM 1:232.
36. Babin 1679, 40; the sentence continues with "il est vray que les Carthesiens ne veulent pas se servir de ce mot d'*infiny,* qui seroit trop odieux, mais seulement de celuy d'*indefiny* qui est la même chose, et qui n'ajoute qu'une seule syllabe à tout ce que nous disons de l'*infiny.*"
37. Ariew 1994a, 6; trans. in ACS 259.
38. Grange 1682, vol. 1, chap. 28: "De la nature du lieu et du vide que le monde est infini, qu'il n'y a point d'espaces vuide au de-la des cieux, et que plusieurs mondes sont impossibles"; Vincent 1677, 69: "An mundus sit indefinite extensus"; Duhamel 1705, 5:16: "Cartesio possibilis non est alter mundus, quia noster mundus est infinite, vel, ut loquitur, indefinite extensus."
39. AT 8:52; CSM 1:232.
40. Ariew 1994a, 4; trans. in ACS 257.
41. Ariew 1994a, 6; trans. in ACS 259.
42. Grange 1682, 1:6. See also Vincent 1677, 75; Duhamel 1705, 5:16.

A sixth, closely related Cartesian principle, also connected with possible limitations of God's omnipotence, can be added to the list of frequently criticized sentences. The relevant Cartesian principles are *Principles* II, articles 16–18, summarized by the proposition "It is a contradiction to suppose that there is such a thing as a vacuum."[43] On this issue, the Oratorians and Jesuits affirmed that "the void is not impossible";[44] the Jesuits condemned the proposition "The compenetration of bodies properly speaking and place void of all bodies imply a contradiction";[45] other Scholastics discussed the issue amply.

This problem arises because seventeenth-century Scholastics deviated considerably from Aristotelian doctrine with respect to the void and motion in the void.[46] Aristotle denied the existence of the void. He argued that the void is impossible, if it is thought to be a place with nothing in it, that is, a location actually existing apart from any occupying body. Further, he concluded against the atomists that motion is impossible in the void, using an argument deriving from his principles of motion.[47] Most Scholastics attempted to soften these arguments, not so as to accept the existence of the void, but to accept its possibility, that is, to argue that God could create a void. Although attacks on Aristotle's views about the void preceded the condemnations of various propositions in 1277, they gained theological inspiration from them.[48] By the seventeenth century, the standard scholastic position was that nature cannot abhor a vacuum because its parts are connected and influence each other,[49] but that God's ability to produce a void—for example by annihilating the sphere of fire or air and not substituting another body for it—cannot be denied.[50] Assuming that God chose to create a void, would there be motion in that void? Scholastics argued that the reasons normally given for the denial of motion in the void would not hold then, because the situation would not be a natural one. Thus, they did not find it difficult to argue against Aristotle on these points. René de Ceriziers's chapters on the void, from his *Le philosophe français*, are typical of late Scholastics discussions:

> Aristotle teaches, in the fourth book of his *Physics*, that motion in the void would be instantaneous because he assumes that the duration of that motion

---

43. AT 8:75–77; CSM 1: 229–31.
44. Ariew 1994a, 4; trans. in ACS 257.
45. Ariew 1994a, 6; trans. in ACS 259.
46. See Chapter 6.
47. *Physics* 215b 21–22.
48. See Duhem 1985, chaps. 9–10.
49. For example, Ceriziers 1643, 2:94–95.
50. Ceriziers 1643, 2:96.

arises only from the resistance of the space. But who does not see that the motion arises also from the quality that produces it, from the succession of its parts, and the distance of its terms? . . . We are led to believe that the Philosopher denied motion in the void against the ancients only because they did not posit any other cause of its duration than the resistance of the medium. From this one could derive the absurdity that a feather would fall as fast in the void as the grindstone of a mill, if it is true that the weight of a body and the distance of its terms cannot be considered.[51]

The scholastic denial of the void is therefore less categorical than its Peripatetic counterpart.

Descartes actually hardened the position, looking more like Aristotle than the Scholastics. He argued for the impossibility of empty space, both in and out of the world. Thinking of a vessel, its concave shape, and the extension that must be contained in this concavity, he asserted: "it would be as contradictory of us to conceive of a mountain without a valley, as to conceive of this concavity without the extension contained in it, or of this extension without an extended substance."[52] In fact, he argued that if God were to remove the body contained in that vessel and did not allow anything else to take its place, the sides of the vessel would thereby become contiguous. Scholastics such as Jean Duhamel took on Descartes's actual argument: "God can absolutely destroy the bodies presently between the heavens and earth, having produced them and conserving them freely . . . God could put a third body between them without displacing them . . . and, as a consequence, heaven and earth would not be touching truly and effectively."[53]

---

51. Ceriziers 1643, 2:97–98. Compare with Léonard Marandé's remarkably similar discussion from *Abrégé curieux de toute la philosophie:*

Ceux qui sont persuadez que la succession necessaire pour mouvoir et porter un corps d'un lieu en un autre, procedoit de la seule resistance de l'air, qui comme une grande mer dans laquelle nous flottons ainsi que les poissons, ne permet pas qu'un corps solide dans un seul moment se puisse acheminer d'une extremité à une autre; ont estimé qu'il ne se pourrait faire de mouvement dans un espace vuide, supposé que la nature en cecy se voulut accorder avec notre hypothese et nous fournir de quoy en faire l'experience. Aristote mesme a esté de cette opinion.

Mais parce que la succession requise pour le mouvement ne procede pas seulement de la resistance de l'air; mais aussi de la distance et de l'esloignement qui se trouve entre l'une et l'autre de ces extremitez supposées, de là vient que supposé le vuide dans la nature, rien ne pourroit empescher le mouvement d'un corps d'un lieu à un autre. (Marandé 1642, 254–55)

52. *Principles* II, art. 18.
53. Duhamel 1692, chap. 4, "Si le vide des philosophes est impossible," p. 202; 1705, 3:203, "Vacuum divinitus possibile est." Cf. also Frassen 1668, 372, "Cartesius contendit, non solum nullum vacuum existere; sec nec etiam divinitus esse possibile," p. 372; Grange 1682, chap. 30, "Si le vide est possible," pp. 410–17; and Vincent 1677, "De vacuo philosophico," p. 63.

What is curious about all this is the feeling of déjà-vu for anyone with the slightest knowledge of the history of condemnations. Most of the difficulties with Cartesianism in the seventeenth century were previously difficulties with Aristotelianism in the thirteenth. Among the propositions condemned at the University of Paris in 1277 were some that were seen as threatening to the Eucharist; prohibited, for example, were "To make an accident exist without a subject has the nature of an impossibility implying a contradiction" and "God cannot make an accident exist without subject or make more than one dimension exist simultaneously."[54] Also condemned in 1277 were numerous propositions thought to infringe upon God's absolute omnipotence, such as "The first cause cannot make more than one world" and "God could not move the heavens in a straight line, the reason being that he would leave a vacuum,"[55] the latter proposition being widely interpreted as a prohibition of the impossibility of a void.

Perhaps Descartes and his contemporaries were unaware of the long history of these prohibitions. The condemnation of 1277 did not seem to be known at the time. There was some consciousness about the issue of condemnation, but it seems to have begun in the 1650s with Jean de Launoy's *De varia Aristotelis fortuna*. Thirteen propositions (out of 219) from 1277 were mentioned in the 1705 treatise by Jean Duhamel, *Quaedam recentiorum philosophorum ac praesertim Cartesii propositiones damnatae ac prohibitae*. The whole condemnation finally appeared in one of the three massive volumes of Charles Duplessis d'Argentré, *Collectio judiciorum de novis erroribus*, published in 1736. However, the interest in these matters clearly stretched back before the 1650s; a sensitivity to the issue of previous historical condemnations of Aristotle's works by the Church was already displayed in Mersenne's *La verité des sciences*, published in 1625.[56]

In any case, there was plenty of evidence, before the advent of Cartesianism, to understand that some of its assertions would be censored. In 1624 the University of Paris and the Parlement prohibited the denial of substantial forms by some anti-Aristotelians on the grounds that holding an atomist philosophy would have been inconsistent with giving an intelligible explanation of transubstantiation.[57] The Cartesians and anti-Aristotelians were not singled out in this respect. It was a common tactic at the start of

---

54. See propositions 196–99 (originally 138–40) in Mandonnet 1908, 175–91.
55. Propositions 27 (34) and 66 (49) respectively. Also proposition 190 (201): "That he who generates the world in its totality posits a vacuum, because place necessarily precedes that which is generated in it; and so before the generation of the world that would have been a place with nothing in it, which is a vacuum." Both in Mandonnet 1908, 175–91.
56. Mersenne 1625, 110–12.
57. For more on what is clearly an important event in the first half of the seventeenth century, see Garber 1988, 1994.

the seventeenth century to claim that a particular philosophical view was not able to accommodate the Eucharist. For example, Scipion Dupleix argued, in his *Physique* (1603), that Thomists cannot explain the Eucharist if they deny that matter can be without form;[58] similarly, he argued that, supernaturally, two bodies can be in the same place, given the Eucharist.[59] Even the possibility of a void was argued on the model of transubstantiation. Théophraste Bouju, in his *Corps de Philosophie,* argued for the *impossibility* of internal place or space to be void of all bodies: "Such a space being a quantity, it is impossible that it can be without body, as much as a quantity is an accident which requires a body in which to inhere, without which it cannot exist,"[60] but he excepted God in his absolute power from this impossibility, on the model of the Eucharist in which God gives subsistence to quantity.[61]

In 1624, Gassendi published his *Exercitationes paradoxicae adversus Aristoteleos,* book 1. He had already written book 2, about Aristotle's logic, and had sketched out books 3 to 7. Apparently, on Mersenne's advice, he forsook the whole project and did not even publish book 2. There, Gassendi had accepted the seemingly innocuous doctrine that "the essence of quantity is nothing but its external extension."[62] As a result, he felt compelled to point out that his doctrine had negative consequences for the sacrament of the Eucharist and to take steps to reaffirm his orthodoxy: "To continue, let us now turn our attention to the famous difficulty concerning the essence of quantity. Our philosophers explain it so well that nothing could be more obscure, though nothing would seem to be more obvious than quantity. However, I must confess that the mystery of the Eucharist, as our faith conceives it, may cause some difficulty in this matter."[63] It seems clear that anyone with an understanding of seventeenth-century philosophy and theology would have appreciated the conflicts between these two domains, as did Gassendi.

Hence, what is needed is a bit of comparative history about the condemned Cartesian doctrines in relation to contemporary scholastic teach-

---

58. Dupleix 1990, 131–32.
59. Dupleix 1990, 261–62.
60. Bouju 1614, 1:468–69, chap. 15, "Comment le lieu environnant peut et ne peut estre vide, par la puissance absolue de Dieu." See also Chapter 7.
61. Bouju 1614, 1:468–69.
62. Gassendi 1624, II, exer. 3, art 10; "Quantitatis essentiam esse extensionem externam." Let us recall that Mersenne also advised Descartes not to publish the end of *Replies* IV to Arnauld, dealing with the Eucharist. Descartes accepted his advice for the first edition (Paris, 1641), but he published the full text a year later in the second edition (Amsterdam, 1642).
63. Gassendi 1624, II, exer. 3, art. 11: "Species Eucharisticas non item fore Fides nos Orthodoxa docet."

ing. I discuss Descartes's definition of substance (including extension as an essential attribute of matter) and his rejection of substantial forms or real accidents in relation to seventeenth-century scholastic doctrines in Chapters 4, 7, and 9. Here I wish to examine the more properly cosmological question, that is, whether the condemnation of *Principles* II, articles 21–22, would be warranted in the light of late scholastic doctrines about the extension and plurality of worlds.

The question of the infinity or finitude of the world depends on the resolution of the question of the possibility of infinity itself, a complex topic. Aristotle's doctrine on infinity may be summarized as follows:[64] there are two infinites to be considered, infinite by addition and infinite by division. But since, when we say that something is infinite, the "is" in that sentence means either what potentially is or what actually is, there are four possibilities to be considered, namely, potential infinite by division and addition, and actual infinite by division and addition. Aristotle denies actual infinities, thereby rejecting both the actual infinitely large and the actual infinitely small. His doctrine about potential infinity is more complex. He wishes to affirm the existence of the potentially infinite by division in magnitude and number, while denying the potentially infinite by addition in magnitude, except in the case where one is adding a part determined by a ratio, instead of keeping the parts equal. However, there is a problem with accepting the existence of the potentially infinite while denying the actually infinite. Generally for Aristotle, what is potential will be actual. This seems to license the inference from the existence of the potential infinite to the actual infinite. The phrase "potential existence" is thus ambiguous. Using Aristotle's example, when we speak of the potential existence of a statue, we mean that there will be an actual statue. It is not so with the infinite. There will not be an actual infinite. "The word 'is' has many senses, and we may say that the infinite 'is' in the sense in which we say 'it is day' or 'it is the games,' because one thing after another is always coming into existence."[65] There are then at least two senses of "potential" according to Aristotle. One sense, which the potential infinite shares with the Olympic games and things whose being is not like that of substance, consists in a process of coming to be and passing away, a process which is finite at every stage but always different. The Olympic games are potential both in the sense that their being consists in a process and in the sense that they *may* occur. It is only in the latter sense that when a state is potential, there will be an actual state. That is how Aristotle can affirm potential infinities such

---

64. Aristotle 1910–62, *Physics* III, chap. 4–8.
65. Aristotle 1910–62, *Physics* III, chap. 6 (206a 18–24).

as the infinite in time, in the generations of man, in the division of magnitudes, and in numbers, while denying the actual infinite. But when Aristotle denies the potential infinite in magnitude by the addition of equal parts, he does so by asserting that "there is no infinite in the direction of increase. For the size which it can potentially be it can actually be. Hence, since no sensible magnitude is infinite, it is impossible to exceed every assigned magnitude; for if it were possible, there would be something bigger than the heavens."[66] Thus, Aristotle's physical world is finite and cannot grow, but in that world magnitude is continuous (or indefinitely divisible) and time and generation are unending (or extendible indefinitely).

The standard scholastic terminology for dealing with the problems of infinity was imported from logic. Logicians distinguished between categorematic terms and syncategorematic terms, or terms that have a signification by themselves, and terms that do not (cosignificative terms). Examples of the first kind are substantival names and verbs, and examples of the second kind are adjectives, adverbs, conjunctions, and prepositions.[67] The distinction is applied to infinity to yield both a categorematic and syncategorematic infinite: "The phrase 'infinitely many' is both syncategorematic and categorematic, for it can indicate an infinite plurality belonging to its substance either absolutely or in respect to its predicate."[68] One can then define the two kinds of infinite separately; syncategorematic infinite may be defined as "for any number or magnitude there is a greater" and categorematic infinite as "greater than any number or magnitude, no matter how great."[69] With the distinction one can solve logical puzzles, since it may be true that something is infinite, taken syncategorematically, and false that something is infinite, taken categorematically.[70] It also enables one to ask separately whether there are syncategorematic and categorematic infinites in nature, without worrying about potentialities. Naturally, various Scholastics took differing views with respect to the existence of various infinites, and often disagreed with Aristotle's doctrines. It is not difficult to see why this should be so, given that portions of Aristotle's doctrine about infinity are clearly in conflict with the conception of an absolutely omnipotent God who is a creator. The standard doctrine (or correction of Aristotle) was the denial of the categorematic infinite (in number and magnitude) and ac-

---

66. Aristotle 1910–62, *Physics* III, chap. 7 (207b 16–21).
67. A list of the syncategorematic terms would normally include: every, whole, both, of every sort, no, nothing, neither, but, alone, only, is, not, necessarily, contingently, begins, ceases, if, unless, but that, and infinitely many.
68. William of Sherwood 1968, 41.
69. See, for example, Rimini 1522, II, fol. 35, col. b. See also Duhem 1985, chaps. 1–3.
70. For example: "I would agree with this [syncategorematic] proposition: along all the parts, a spiral line is drawn; and I would not agree with this [categorematic] proposition: a spiral line is drawn along all the parts," Buridan 1509, fol. 59, col. c.

ceptance of the syncategorematic infinite (in number and magnitude); this would be equivalent to denying actual infinites but accepting all potential infinites. Of course, there were some thinkers, notably Gregory of Rimini and Albert of Saxony, who argued that God could create a categorematic infinite in nature.

The seventeenth-century doctrines generally conflated syncategorematic infinite with potential infinite and categorematic infinite with actual infinite, but denied the inference from syncategorematic infinite to categorematic infinite.[71] There followed a denial of infinity in act. However, seventeenth-century Scholastics were also careful to state that others argued that God could create a categorematic infinite. The condemnation of various propositions in 1277 influenced the discussions of the possibility of syncategorematic and categorematic infinites in nature. Among the condemned propositions was number 34: "That the first cause cannot make more than one world."[72] This proposition challenged directly the Aristotelian doctrines of the singularity of the universe and the impossibility of the potential infinitely large in magnitude. It also suggested that one should be careful when denying the actual infinitely large. Here is a typical Thomistic-leaning pessimistic discussion:

> There is no real being which is not either finite or infinite. The infinite can be considered in two ways: simple and in some kind. Simple infinite is that which is not limited by any bounds nor any kind of manner of being, but which contains in itself eminently all degrees of being, all those that can give perfection without depending on any other thing. This kind of infinite is suitable for God alone, who is infinite in entity, essence, perfection, and duration, the reason being that he has existence by himself and all others have it from him. . . .
>
> The infinite in some kind (if it is found in nature) would be something finite and determined to a certain genus according to its essence, but infinite in quantity of extension, or discreteness, or perfection, or duration; such an infinite can only be considered with respect to proper or improper quantity: for example, if some continuous quantitative thing not limited in its extension were given, it would be infinite along the essence and being of quantity, that is, in extension, but it would not be infinite simply, because it would be finite and limited under the genus quantity. Similarly, if it were so white as not to have any limit in its whiteness, it would be intensively so according to its proper essence and perfection of whiteness, for this whiteness would be infinite, and yet it would not be infinite simply, to the extent that it would be limited under

---

71. Goclenius (1613, 237), for example, states, "Syncategorematice: Potentia, mentali abstractione, ut Zabarella loquitur. Ab infinito in potentia, ad infinitum actu nulla est consecutio. Categorematice: Actu. Haec immensitas non potest communicari ulli creaturae."

72. Proposition 27 in Mandonnet 1908, 175–91.

the genus of infinite quality.... If there were a body infinite along all dimensions, it would be alone in the universe, for it would occupy all place—that which we know to be false, there being several bodies of differing nature in the universe. As for discontinuous quantity, there is no such infinite either, for if it were so, it would be a certain numerable or innumerable multitude....

The infinite in some kind, of which we have just spoken, does not occur only in act, but also in potentiality, for if it occurred in passive potentiality it could be reduced to act, given that nature does nothing in vain. If it occurred in objective or active potentiality, that is, in the potentiality of the agent, God at least could produce it, since it would not contain any contradiction. But this cannot be, given that God is not able to do more than the infinite, and thus he could not do anything after having accomplished it; therefore, his power would be less than it was before—in fact it would be nothing at all—since he could not do anything after the infinite. This is very absurd, given that his power is neither diminished nor augmented, nor can it be spent by the production of things it makes outside of itself.... That is why one says that God has the infinity of power but not the power of infinity, meaning that while his power is infinite, he cannot make any infinite thing distinct from his essence.[73]

Others were a bit more positive about the possibility of an actual infinite. In his *Physics,* Toletus treats such topics as the categorematic infinite, division into proportional parts, and the question whether a body can be actually infinite,[74] but he affirms a generally conservative position. On the other hand, he does refer his readers to Albert of Saxony's more daring position: "Alber. Saxo. hoc lib. q. 9."[75] Roughly the same can be said about the Coimbrans[76] and Abra de Raconis, except that de Raconis gives specific citations to William of Ockham and Gregory of Rimini: "Prior est Ochami in 2. qu. 8 & quodlibeto 2. q. 5 Greg. Ariminensis in 1. dist. 43. q. 4 & aliorum per divinam potentiam infinitum actu categorematicum posse creari."[77] According to L. W. B. Brockliss, there was a Schoolman—du Chevreul, Professor at Paris in the 1620s and 1630s—who taught that Aquinas was wrong to deny that God could create an infinite body.[78]

Eustachius a Sancto Paulo's doctrine seems to differ significantly from the standard view, so it is worth detailing. It looks as if Eustachius thinks of syncategorematic infinite as a species of infinite in act. In his *Physics,* tract. 3, question 5, "What and in what way is something infinite," Eustachius divides the infinite into infinite in actuality and potential infinite. He then di-

---

73. Bouju 1614, 1:275–78.
74. Toletus 1589, III, quaest. v–vii, fol. 100, col. a–fol. 103, col. d.
75. Toletus 1589, fol. 103, col. a.
76. Conimbricenses 1592, *Physics,* vol.1, col. 509–40, especially col. 524.
77. Raconis 1651, pars 3, p. 194. De Raconis is incorrect in attributing the categorical infinite in act to Ockham.
78. Brockliss 1987, 338. (But see below for de Ceriziers's doctrine.)

vides the former into categorematic actual infinite and syncategorematic actual infinite, depending upon whether all the parts of a given infinite are actually separated or not. Infinites whose parts are not all in actuality are of three kinds: infinite in succession, addition, and subtraction.[79]

Eustachius does think that the continuum is divisible into infinite parts. But, in the final analysis, his doctrinal deviations from the standard view are more cosmetic than real. Eustachius argues that the continuum is not divisible by equal magnitudes; it is divisible by equal proportional parts (or by parts whose magnitudes diminish by halves). Thus it is infinitely divisible successively, and not simultaneously. The continuum is divisible to infinity not so that there can exist simultaneously actually separated infinite parts, but so that one can progress in the division:

> If you object that it follows that if one has to posit an actual infinity in nature, it would follow that either one can divide a continuum into infinite parts or those parts in the continuum would not be actually infinite, we reply, *infinity in act* can be conceived in two ways: one, properly speaking, in which all the parts are actually separated and distinct from one another, which is called categorematic infinite; the other in truth, improperly speaking, whose parts are not actually separated from one another, but are said to be communicating with one another, in which the smaller are contained in the larger, which is called syncategorematic infinite. Thus a continuum can be divided to infinity and it does not follow that we have to hold an actual infinity, properly speaking, but only an infinite in act in the second way, improperly speaking. From this it is to be understood that all parts of the continuum are actually in the continuum, not however actually infinite categorematically and properly, but syncategorematically and improperly.[80]

Eustachius is playing verbal games with "actual infinity" and "syncategorematic infinite." He does not really hold that syncategorematic infinites are, properly speaking, actual infinites. In fact, he reaffirms that "only the actual categorematic infinite is truly and properly infinite.... Thus the actual syncategorematic infinite is not properly an infinite in act... it is to be called potential infinite."[81] And he rejoins the standard doctrine. He is even careful to look as if he is upholding God's absolute omnipotence when denying him the power to create a categorematic infinite: "There is no actual categorematic infinite, not because it is repugnant to God's power, but because nature cannot suffer it."[82]

---

79. Eustachius 1629, Physics, tract. 3, quaest. 5, "Quid et quotuplex sit infinitum," p. 54.
80. Eustachius 1629, Physics, tract. 3, quaest. 4, "An continuum sit divisible in infinitum," p. 53.
81. Eustachius 1629, Physics, tract. 3, quaest. 5, p. 54.
82. Eustachius 1629, Physics, tract. 3, quaest. 7, "An detur aut falsum dari possit infinitum," p. 56.

The seventeenth-century scholastic answers to the question of whether God can produce an infinite body ran the gamut from Scipion Dupleix's negative answer—there is no infinite body in nature and it is not repugnant to God's power not to be able to produce one[83]—to René de Ceriziers's positive answer: "Although we refuse Nature the power of producing the infinite, we should not refuse it to its Author. Can he not make everything he can in this moment, for example, can he not make all the men he can produce? If so, their multitude will either be finite or infinite. Let us say that it is finite; that would be to limit God's power. To grant that it is infinite is to agree with my opinion."[84] De Ceriziers proceeded to reject all arguments claiming that an actually infinite world would be impossible.[85]

Thus there were precedents for Descartes's doctrine of the indefinite extension of the universe. For Descartes, God is the only being in whose perfections one notices no limits,[86] and one can see that he is greater than the world,[87] so that the world cannot be called infinite.[88] But it conflicts with one's conception for the world to be finite or bounded.[89] Hence one calls it indefinite.[90] Descartes allegedly said to Franz Burman that the distinction between the infinite and the indefinite "is the author's [Descartes's] invention."[91] Moreover, defending himself against the claim that his hypothesis of an infinite would do harm to Christianity,[92] Descartes responded that "Cardinal Cusa and several other doctors have supposed the world to be infinite without ever being censured by the Church," and that his own "opinion is not as difficult to accept as theirs."[93] Oddly enough, Nicholas of Cusa was the figure whose doctrine about the extension of the world was closest to Descartes's. In his *De docta ignorantia,* Cusa asserted that only the absolute maximum is infinite, for it alone is everything it can be. The universe does not include everything existing outside God, but it is not God; therefore it is not positively infinite. However, there is no term limiting the universe, so that we can call it infinite, if we take the word "infinite" in a privative sense signifying an absence of limit. More exactly, we might say

---

83. Dupleix 1990, 287–91. See also Goudin 1864 [1668], 2:460–82, for the argument that God alone is infinite in act and that infinity in act is impossible for any creature, magnitude, or multitude.
84. Ceriziers 1643, 116.
85. Ceriziers 1643, 116–19.
86. *Principles* I, art. 27.
87. AT 5:345.
88. AT 5:52.
89. AT 5:345.
90. AT 5:345; *Principles* II, art. 21; AT 11:656. For the extension of matter called indefinite, see AT 5:274–75, 7:112–14. For more on Descartes and the infinite, see Ariew 1987a.
91. AT 5:167.
92. AT 5:20–21.
93. AT 5:51.

that the universe is neither finite nor infinite.[94] For Cusa the universe is indeterminate in both senses of the word: it does not have a boundary (it is not terminated), and it lacks precision (it cannot be determined by us).[95] Hence, Cusa's doctrine is remarkably like Descartes's. Still, Descartes was right in thinking that his opinion should not have caused any difficulty for Christians.

Although there are precedents for thinking the world infinite or indefinite in extent in scholastic philosophy, I have not, however, been able to find a late scholastic philosopher who argued that there cannot be a plurality of worlds. The closest anyone comes to such an assertion is the argument that there would a unity to the world if there were only one; but even that argument is couched in a language that allows for the possibility of plural worlds: "Is it not likely that, being able to create an infinity of worlds, he produced only our own in order to mark the unity of the Creator in the unity of the work?"[96] Indeed, the most conservative thinkers take it for granted that God, who created this world, could create another.[97] Thus, Descartes is truly out on a limb by himself on this final issue.

We are left with one last puzzle, which we will leave unresolved for now. Given the above, how could Descartes have asserted sincerely in 1637 that he is able to thank God for the fact that the opinions that have seemed to him most true in physics, when considering natural causes, have always been those which agree best of all with the mysteries of religion?[98]

---

94. *De docta ignorantia*, II, chap. 1, in Cusa 1962 [1514], vol. 1, fol. 13.

95. *De docta ignorantia*, II, chap. 1, in Cusa 1962 [1514], vol. 1, fol. 13v. See also *Idiota de mente*, chaps. 2 and 3, in Cusa 1962 [1514], vol. 1, fols. 81v–83v.

96. Ceriziers 1643, 121.

97. For example, Bouju does not treat the issue directly, but when discussing what there is above the first heaven, he claims, in good Aristotelian fashion, that there is nothing: no body can have its natural place beyond the first heaven; there are no surrounding places for a body to be contained there, and thus no void nor a real space. The only thing one can say, according to Bouju, is that there is an imaginary space above the first heaven "which is nothing else than its non-repugnance toward being the situation of a body, if God wanted to create one there" (1614, 1:374). For the background to these issues, see Duhem 1985, chaps. 11–13.

98. AT 1:456; CSMK 75. But see Chapter 7 for what Descartes may have had in mind with that statement.

[9]

Cartesians, Gassendists, and Censorship

During the second half of the seventeenth century, Cartesians suffered a series of condemnations issuing from various authorities in France. The condemnations aimed at several fundamental propositions of corpuscularianism and the mechanical philosophy, such as the denial of substantial forms and real qualities. Also condemned was the Cartesian theory of matter and place, that is, the doctrine that extension is the principal attribute of matter, and some of its consequences, namely, the indefinite extension of the world and the rejection of the void.[1] These propositions were deemed incompatible with some of the mysteries of Catholicism and thus disruptive to the faith and to the public order. Moreover, it was well known—and commonly argued at the time—that Gassendism, the revival of Epicurean atomism, was for similar reasons equally incompatible with the same mysteries, as was the Gassendist doctrine of matter and space. But the Gassendists did not suffer the same fate as Cartesians; they were not condemned officially. But why were they not? From the point of view of the authorities, there should not be anything to choose between René Descartes and his followers and Pierre Gassendi and his followers. Nor if social causes are to be invoked to account for the asymmetry of condemnations, will they be found at the surface. After all, Descartes and Gassendi, two Frenchmen from the provinces, traveled in the same circles—they were both friends and correspondents of Marin Mersenne—though Gassendi's circle also included the *libertins érudits*. Unlike Descartes, Gassendi actually published an anti-Aristotelian work as early as 1624. Moreover, although Gassendi was a priest and professor of

1. See Chapters 7 and 8.

philosophy at the Collège d'Aix, he was removed from his position when the Jesuits took over the college in 1623. Descartes, on the other hand, was taught by the Jesuits of La Flèche for more than eight years, and, in spite of his seemingly harsh pronouncements about his education in the *Discourse on Method* and his war with the Jesuit Pierre Bourdin in the 1640s, he generally maintained good relations with them throughout his adult life.[2] Later in life, in 1645, Gassendi was nominated professor of mathematics at the Collège Royal, but because of ill-health, he taught there only a year. One cannot underestimate Gassendi's fame at the time, but his royal connection would seem to have been counterbalanced by the handsome pension bestowed on Descartes by the king in 1646.

In the second half of the seventeenth century Descartes and Gassendi gained followers from all walks of life. Perhaps Descartes's followers were predominantly Jansenists and Oratorians and Gassendi's were Jesuits and others, but those facts themselves, if they are facts at all, would require much exposition and explanation: what was there about Descartes and Cartesianism that might have attracted Jansenists and Oratorians but not Jesuits?[3] Similar questions may be asked about Gassendi and the Gassendists. Superficially, there would seem to be little to choose between the two challengers to the dominant Aristotelian intellectual tradition. So let us examine the two cases in greater detail.

The Cartesians

With the Jesuits conspiring in the background, as was allegedly their fashion, the Catholic church in 1663 put Descartes's works on the *Index of Prohibited Books* with the notation *donec corrigantur*—"until corrected." Descartes had been dead for thirteen years. It was not likely that he would correct his works; so the prohibition was effective. Antoine Arnauld, complaining about the placing of Descartes's works on the *Index,* wrote: "it is true that the prohibition was only *donec corrigantur*. But that could not be done. Since there is no indication of what is to be corrected, it is the same thing as if the books were prohibited absolutely."[4] Thus, the prohibition was as effective as the Catholic Church could make it. However, the Church

---

2. See Chapters 1 and 7.
3. It is easy to overstate the affinity between Port-Royal and Descartes. Steven Nadler reminds us that "the majority of those connected with Port-Royal professed an open hostility towards Cartesian philosophy. In fact, Arnauld appears to be one of the very few Port-Royalists of his generation to have had any sympathy towards this new philosophy," Nadler 1989, 18. One can also find Cartesian sympathizers among the Jesuits both during Descartes's life and after. Thus, pronouncements about Jansenist Cartesians must be handled carefully.
4. Letter 830 to du Vaucel, Arnauld 1775, 3:398.

did not have any authority in the Protestant world; it did not even have authority in Catholic countries such as France, where censorship was the domain of temporal powers: the king and individual universities.⁵ Moreover, Cartesianism seems to have made great inroads into French universities in the 1650s and 1660s, spreading from private lectures and salons into university teaching and student writings, even though candidates for chairs in philosophy were required to deny the so-called new philosophy and to fight against Descartes.⁶ The situation was about to change in the 1670s. In 1671, the archbishop of Paris, François de Harlay, published the following verbal decree from the king:

> The King, having learned that certain opinions that the faculty of theology had once censored, and that the parlement had prohibited from teaching and from publishing, are now being disseminated, not only in the University, but also in the rest of this city and in certain parts of the kingdom, either by strangers, or by people internally, and wishing to prevent the course of this opinion that could bring some confusion in the explanation of our mysteries, pushed by his zeal and his ordinary piety, has commanded me to tell you of his intentions. The King exhorts you, sirs, to make it so that no other doctrine than the one brought forth by the rules and statutes of the University is taught in the Universities and put into theses, and leaves you to your prudent and wise conduct to take the necessary path for this.⁷

The reference to "certain opinions that the faculty of theology had once censored" was, interestingly enough, a reference to a condemnation of atomism in 1624, in which some in Descartes's circle, Mersenne and Jean-Baptiste Morin, for example, had played a role.⁸ That condemnation was being used against Cartesianism almost five decades later. The "confusion in the explanation of the mysteries" was also a reference to the same episode. One of the reasons given for the condemnation was that holding an atomist philosophy would have been inconsistent with giving an intelligible explanation of transubstantiation.⁹ It was clear that the scholastic metaphysics of matter and form was well suited to explain the mystery. While the religious and political authorities might have had many objections to Cartesianism (or to any atomist or corpuscularian philosophy) the

---

5. See McClaughlin 1979.
6. Bouillier 1868, 1:468.
7. Bouillier 1868, 1:469.
8. Both Mersenne and Morin applauded the condemnation. For Mersenne's description of theses, see Chapter 4. For an analysis of the event, see Garber 1988, 1994.
9. See Armogathe 1977. The loci for Descartes's view are the end of *Replies* IV, *Replies* VI, section 7, and the *Letters to Mesland* (AT 4:163–72, 346–48). It was not out of the ordinary for the king to attempt to protect the mysteries of the faith. Neither were Cartesians and anti-Aristotelians singled out in this respect. See Chapters 7 and 8.

king's edict had the effect of focusing the criticism on the Descartes's account of transubstantiation, something the Faculty of Theology of the University of Paris expressly censored, the Parlement prohibited, and the king cautioned about. In any case, the king's exhortation was a serious threat to Cartesianism in France; various universities—Angers, Caen, Paris—followed with attempts to carry out the king's wishes.

We have a first-hand account of the subsequent events at Angers in a journal kept by François Babin, doctor of the Faculty of Theology and financial officer of the university there—and someone horrified by the attitudes of the Cartesians.

> Young people are no longer taught anything other than to rid themselves of their childhood prejudices and to doubt all things—including whether they themselves exist in the world. They are taught that the soul is a substance whose essence is always to think something; that children think from the time they are in their mothers' bellies, and that when they grow up they have less need of teachers who would teach them what they have never known than of coaches who would have them recall in their minds the ancient ideas of all things, which were created with them. It is no longer fashionable to believe that fire is hot, that marble is hard, that animate bodies sense pain. These truths are too ancient for those who love novelty. Some of them assert that animals are only machines and puppets without motion, without life, and without sensation; that there are no substantial forms other than rational soul; and by completely contrary principles . . . others teach that the souls of animals are immortal, spiritual, and created directly by God, as are those of men.[10]

It is clear that, for Babin, something had gone terribly wrong. He continued his observations, moving from pedagogical and epistemic to metaphysical and theological problems, and ultimately to political ones:

> The Cartesians assert that accidents are not really distinct from substance; that it would be well to guard oneself from attributing some knowledge or certainty to the testimony of our senses. . . . They make the essence of all bodies consist in local extension, without worrying that Christ's body does not better accommodate their principles and our mysteries; they teach that something does not stop being true in philosophy even though faith and the Catholic religion teach us the contrary—as if the Christian and the philosopher could have been two distinct things. Their boldness is so criminal that it attacks God's power, enclosing him within the limits and the sphere of things he has made,

---

10. Babin 1679, 2.

as if creating from nothing would have exhausted his omnipotence. Their doctrine is yet more harmful to sovereigns and monarchs, and tends toward the reversal of the political and civil state.[11]

Babin seems to have been right. The Cartesians were so out of control that, far from heeding the edict of the king and of the archbishop of Paris, they were making a mockery of it. Upon hearing the king's decree, Boileau, Racine, and others had counterattacked with their own decree. If the king and his henchmen were going to condemn Cartesianism, the Cartesians were going to condemn them to their fate for having supported Aristotle.

The satire traveled even to Angers. Babin reproduced a version of it; his journal entry was introduced with the following comment: "We produce this piece here in order to show that the innovators use all their wit and industry in order to evade and translate into ridicule the powers that fight against them; and that they do not fail to use mockery, caricatures, or jokes, in order to validate their decried opinions, wishing by that means to dazzle the common minds by the effect of a false light and to persuade the rabble that reason, truth, knowledge, and good sense are theirs alone."[12] The "arrêt burlesque," as it was known, with its mock legalese language, read as follows (after considerable preliminaries):

> The Court having respect to the aforementioned request, has kept and maintained, [and still] keeps and maintains the said Aristotle in the full and peaceful possession and enjoyment of the said schools, and orders that he will always be followed and taught by the regents, doctors, masters of arts and professors of the said University, without their being required to read him, or to know his language and opinions, and with respect to the principles of doctrine, returns them to their notebooks with the injunction to the heart to continue to be the principle of the nerves, and to all the professors of whatever quality, condition, and profession, to believe it to be so, notwithstanding all experiences to the contrary; similarly orders the digestive juices (*chyle*) to go straight to the liver, no longer passing through the aforementioned heart, and to the liver to receive it.... It reestablishes the entities, identities, virtualities, haecceities, and similar Scotist formalities into good repute, ... it banishes Reason to perpetuity from the schools of the aforementioned University, prohibits it from entering there, from troubling or bothering the aforementioned Aristotle.[13]

As might have been expected, the authorities at Angers prevailed. They submitted some professors' writings to examination and found that the au-

---

11. Babin 1679, 2.
12. Babin 1679, 18.
13. Bouillier 1868, 1:471. Babin 1679, 18. Boileau 1747, 3:150–53. Murr 1992, 241–40.

thors were teaching the prohibited propositions; moreover, they also found that the professors allowed their students to publish theses without their being submitted to examination by the rector. Consequently, Fathers Fromentier and Villecroze of the Oratory, professors of philosophy at the College of Angers, were censured. Father L'Amy and his successor, Father Pélaut, were ultimately prohibited from teaching and exiled from Angers.[14]

In Paris, the situation was different. Although the theology faculty formally condemned Cartesianism in 1671 and the faculty of medicine followed suit in 1673, the faculty of arts took no action. Perhaps the "arrêt burlesque" succeeded—at least Boileau thought so.[15] Perhaps also an anonymous treatise entitled "Several Reasons for Preventing the Censure or Condemnation of Descartes' Philosophy" had an effect.[16] The anonymous treatise was most probably written by Arnauld; Victor Cousin, who reprinted the treatise, claimed to have found a manuscript of it dated 1679 and attributed to Arnauld.[17] Cousin also held that the arguments and style of the treatise resembled greatly those of Arnauld. Although it was not reproduced in Arnauld's collected works, it can be found in an eighteenth-century edition of the works of Boileau, right next to the "arrêt burlesque."[18]

In the treatise, Arnauld gave ten reasons against the condemnation of Cartesianism. He intimated that the attack on Descartes was a political ploy of the Jesuits and various enemies of the Jansenists attempting to create troubles for the Jansenists.[19] But he said, even those without such political intentions can cause problems unwittingly, because it is impossible for an edict to change people's opinions and to cause those who do not accept a particular philosophy (such as Aristotle's) to embrace it. By necessity such an edict can only be general and thus cause endless disputes, since everyone can interpret it as they wish. In any case, people's minds are not so flexible that anyone can have the freedom of believing whatever they want. As

---

14. Babin 1679, 35–45. In fact, in 1678, the Oratorians and Jesuits got together at a general congress and agreed *not* to teach Descartes's new philosophy. For the text of the prohibition, see Ariew 1994a, 4; trans. in ACS 256–57.

15. See Boileau 1747, 3:108.

16. *Plusieurs raisons pour empêcher la censure ou la condamnation de la philosophie de Descartes,* Boileau 1747, 3:117–41. Although anonymous, it was known to have been written by someone from Port-Royal.

17. Cousin 1866, 3:303–17.

18. Boileau 1747, 3:108–54.

19. From Arnauld's point of view, the equation Cartesian/Jansenist was unproblematic. The difficulties that Cartesians were having were political, according to Arnauld, the Jesuits acting against their enemies, the Jansenists. Arnauld also argued that since Descartes dedicated his metaphysics to the Faculty of Theology at Paris in 1641, the fact that the faculty had been silent about Descartes's work for thirty years indicated that the attempt at condemnation was a political act (Cousin 1866, 309).

the history of previous condemnations showed, you cannot succeed in requiring people to hold a particular philosophy. When this is tried, the authority of the church is compromised. Arnauld then listed various condemnations of philosophy, pointing to absurdities and contradictions in the prohibitions.

Arnauld based his account on de Launoy's *De varia Aristotelis fortuna*, published in 1653; it was a part of a growing interest during the seventeenth and eighteenth centuries in the history of condemnations, an interest evinced in de Launoy's work, in a 1705 treatise by Jean Duhamel, *Quaedam recentiorum philosophorum ac praesertim Cartesii propositiones damnatae ac prohibitae*, and in the three massive volumes of Duplessis d'Argentré, *Collectio judiciorum de novis erroribus*, published in 1736.

Arnauld's first group of condemnations detailed the thirteenth-century battles between Aristotle and the church. Notwithstanding various church condemnations and prohibitions in 1209, 1215, and 1231, Albertus Magnus and Thomas Aquinas taught and commented upon Aristotle's books. Similarly, in 1264, Aristotle's books on metaphysics and physics were again prohibited by Apostolic authority. Yet two years later students were receiving degrees based on their readings of Aristotle's prohibited books. In contrast, the second group concerned the prohibitions of anti-Aristotelian writings by the church in the sixteenth and seventeenth centuries, from Ramus's criticisms of Aristotelian logic to the anti-Aristotelian atomist opinions of de Clave, Villon, and Bitaud. Arnauld pointedly indicated that the 1624 anti-atomist edict, which forbade any teaching against the approved ancient authors, with capital punishment as penalty, did not prevent Gassendi from publishing the *Exercitationes paradoxicae adversus Aristoteleos* in the very same year.[20] Finishing his history of condemnations, Arnauld referred to the medieval battles between realists and nominalists and to the decree of Louis XI against the nominalists, claiming that the edict cannot be read without being thought ridiculous and a testimony to the narrow-mindedness of the human mind.

20. One should note that as early as 1624, Gassendi had announced his intention of writing against Aristotle's doctrines of space and the void: "Book III is devoted to an exposition of physics. A number of the Aristotelian principles are attacked in it; among other things, it is proved that forms are accidental.... The space of the ancients is recalled from exile and is substituted for Aristotelian place. Void is introduced, or rather reestablished, in nature." (Gassendi 1959, 12–15; trans. in ACS 169). And Gassendi was almost immediately recognized as an opponent of Aristotle's, as can be seen in Frey 1628, chap. 3, "In quo Petrus Gassendus innumera falsissima, et impia Aristotelem protulisse docens cribratur," pp. 37–41; chap. 12, "Patricius, Gassendus et Campanella de infinito, de vacuo, de ideis, de lineis, de Galaxia contra Aristotelem sententientes, reiiciuntur," pp. 59–63; chap. 15, "Patricius et Gassendus impiam omnem et falsam de Deo doctrinam Aristotelicam asserentes, cribrantur," pp. 67–73; and chap. 16, Ramus, Ludius, Patricius Gassendus reiiciuntur, asserentes nullum Peripateticorum usum Dialectices novisse," pp. 73–75.

Arnauld then launched into a discussion of the problem of the explanation of the Eucharist, beginning with the general point that a condemnation of a particular philosophy on the ground that it cannot explain the Eucharist would be harmful to religion. There would be no advantage to the church in having a popular philosophy, embraced by countless Catholics, declared as incompatible with the mystery of the Eucharist. (The act might even be used by Calvinists to infer that, like them, some Catholics did not believe in transubstantiation.) In addition, Aristotle himself was not without difficulty in this respect, according to Arnauld. Aristotle taught that a body cannot be in several places at once. Thus, even Aristotle needed to be reconciled with faith by means of a principle like "it is of the nature of the infinite not to be able to be understood by that which is finite,"[21] a principle which, Arnauld asserted, is needed to reconcile any philosophy with faith. Arnauld then recommended a demarcation between theology and philosophy in order to avoid purely philosophical questions in theology. He cited with approbation such cardinals as Du Perron and Richelieu, who had avoided mixing philosophy with theology and who had accepted the mysteries of Catholicism as matters of faith not needing naturalistic explanation. Arnauld then returned to the condemnation of 1624 against the atomists' rejection of substantial forms (except for rational soul). He referred to two good Catholics who had also rejected substantial forms in various published works: the Jesuit Fabri, in a work dedicated to the general of the Jesuits,[22] and the Minim Maignan, in a work approved by the superiors of his order.[23]

As his final argument, Arnauld pleaded that there was nothing wrong with leaving things untouched, as they had been for many years, and that it was always good to maintain the peace and not to give those who wish to trouble the peace the opportunity to do so.

While Arnauld's treatise was a general counterattack against those who wished to prohibit certain works, and as such was a complex document full of nuances, still there are two items that particularly need to be emphasized. (1) Given the context of the second half of the seventeenth century, that is, given the works of Fabri and Maignan, which had already received church approbation, it would have been difficult to prohibit the new philosophy on the ground that it dispensed with substantial forms; a prohibition simply using the condemnation of atomism in 1624 would have been fairly weak. (2) Arnauld and other Cartesians had adopted the tactic of

---

21. "Quand nous considérons, d'une part, la puissance infinie de Dieu, et de l'autre, la foiblesse de notre raison, le bon sens doit nous faire juger qu'il n'est pas étrange que Dieu puisse faire ce que notre raison ne savoit comprendre" (Arnauld, in Cousin 1866, 3:311).
22. Fabri 1666.
23. Maignan 1653.

linking various philosophies together, arguing that if one must be prohibited, all must be prohibited. In particular, the Cartesians argued that Gassendists were as guilty as, or even more guilty than, Cartesians with respect to some alleged heresies. We see this again in a letter written by Arnauld, in a passage concerning the prohibition of Descartes's works by the censors of Rome:

> I am not surprised by what I am told about Naples, that some young fools have become atheists and Epicureans through the reading of Gassendi's works. It is what one ought to expect, especially when one considers what he has written against Descartes' metaphysics, where he has used all his wit to attack everything powerful Descartes had found to prove the existence of God and the immortality of our souls. Isn't there something to admire in the great judgment of the inquisitors of Rome and the great service they render to the church by their prohibitions? They have allowed a freedom to these young people to read the author who destroys, as much as he can, the most solid proofs of the existence of God and the immortality of our souls (for there are no works of Gassendi on the *Index*), but they did not allow them to read the one who would have persuaded them of these truths, for fear that their minds would be set in the right direction; for the censors of Rome had care to place in their *Index: Renati Descartes Opera sequentia donec corrigantur. De Primà. philosophiâ in quâ Dei existentia, & anima humanae à corpore distinctio demonstratur.*[24]

Some years later, in 1691, the University of Paris finally took formal action and condemned eleven propositions of Descartes, including the propositions "One must rid oneself of all kinds of prejudices and doubt everything before being certain of any knowledge," "One must doubt whether there is a God until one has a clear and distinct knowledge of it," and "We do not know whether God did not create us such that we are always deceived in the very things that appear the clearest."[25] Interestingly, the 1691 condemnation did not make any direct mention of the issue of the alleged incompatibility between Cartesianism and the mysteries of Catholicism. It did not prohibit the proposition at issue in the condemnation of 1624, the denial of substantial forms and real qualities. The closest the condemnation got to this subject was the obscure condemnation of the proposition "As a philosopher, one must not develop fully the unfortunate consequences that an opinion might have for faith, even when the opinion appears incompatible with faith; notwithstanding this, one must stop at that opinion, if it is evident."[26] I should add that this did not prevent the

---

24. Arnauld, letter 830 to du Vaucel (1775, 3:395–98).
25. Argentré 1736, pt. 1, p. 149.
26. D'Argentré 1736, pt. 1, p. 149. The authorities also condemned propositions whose avowal was not usually connected with Descartes. See Ariew 1994a; trans. in ACS 257. Of

scholastic textbook authors—Jean Duhamel, Claude Frassen, Antoine Goudin, Jean-Baptiste de la Grange, and others—from criticizing Descartes's principles and his denial of substantial forms on theological grounds, both before and after 1691.[27]

In any case, the authorities struck at the Cartesian doctrine of matter and extension with a prohibition of the related doctrine that "the matter of bodies is nothing other than their extension, and one cannot be without the other." The denial of this proposition was also connected with the problem of the Eucharist and the interpretation of transubstantiation given at the Council of Trent. It had become the main issue for those who wished to claim that Cartesianism was incompatible with the Eucharist; indeed, it was the battleground chosen by the Père de Valois, a Jesuit of Caen, who wrote a treatise under the alias of Louis de la Ville, entitled *Sentimens de Monsieur Descartes touchant l'essence et les proprietez du corps opposez à la Doctrine de l'Eglise, et conforme aux erreurs de Calvin sur le sujet de l'Eucharistie*. As Pierre Bayle stated about de la Ville's treatise, "It is clear that the Council of Trent has decided not only that the body of Christ is present everywhere there are consecrated hosts, but also that all the parts of his body are interpenetrated. It is clear, because of de la Ville's book, that this decision is absolutely incompatible with the doctrine positing that extension is the whole essence of matter."[28]

The Gassendists

That was also the problem François Bernier, the most ardent defender of Gassendism in the seventeenth century, set out to resolve in his *Eclaircissement sur le livre de M. de la Ville*. Bernier needed to reply to de la Ville because his attacks were directed generally against the new philosophy and thus also against Gassendi.[29] As Bayle explained: "Bernier, so well known because of his travels and the regard the celebrated Mr. Gassendi had for him, and because of the public testimony he gave of his veneration and

---

course, the condemnation of 1691 cannot be considered the necessary consequence of the events of 1671–78; it itself needs to be contextualized. Here again the "arret burlesque" might come to the rescue: in a later edition of it, the line referring to the heroes of the "arret," "Cartistes et Gassendistes," was expanded to include "Pourchochistes et Malebranchistes." It does not take a lot of imagination to conclude that much of the condemnation of 1691 was directed against Edmond Pourchot, the first professor of philosophy at the University of Paris openly espousing Cartesian doctrines; Pourchot taught at the Collège Mazarin from 1690 to 1704. See Brockliss 1987.

27. See, for example, Duhamel 1705; Frassen 1686; Goudin 1868; Grange 1682; Vincent 1677.
28. Bayle 1684, "Avis au lecteur," fol. 7v–8r, unpaginated.
29. Bernier, *Eclaircissement sur le livre de M. de la Ville*, in Bayle 1684, 45–91.

gratitude for so great a Master, fearing the malign influences of the zeal of these people, had secretly published a small treatise (the third piece of this collection), copies of which he distributed in secret to his friends and even to some prelates."[30] The content of the small treatise, however, was not as high-minded as one would have hoped for; according to Bayle, "He lets them do whatever they want to the Cartesians and declares himself very vigorously against some Cartesian doctrines, in order to make his peace more easily; having as many reasons as the Cartesians to fear that he would be accused of heresy with respect to transubstantiation, he does what he can in order to have his innocence known."[31]

Oddly, Bernier, at the same time, was publishing the second edition of the *Abrégé de la philosophie de Gassendi*,[32] and he had just published a small work entitled *Doutes sur quelques-uns des principaux chapitres de son abrégé de la philosophie de Gassendi*,[33] most of which he also inserted as "Doutes sur quelques-uns des principaux chapitres de ce tome [i. e., vol. 2]," in his second edition of the *Abrégé*.[34] In these two places, Bernier seemed to abandon Gassendi's doctrine of an absolute space independent of things, in which all things are contained and succeed each other,[35] in order to identify space and body, as did Descartes.[36] The seeming inconsistency of someone criticizing the Gassendist theory of space and leaning toward the Cartesian doctrine of space while delivering Cartesianism to its scholastic and church opponents on the very same issue so disturbed Francisque Bouillier that he felt the need to chide Bernier some two centuries later; echoing Bayle's accusation, Bouillier wrote: "Bernier is wrong to seek to prove his innocence at the Cartesians' expense, even more so since he himself abandons Gassendi in order to get nearer to Descartes in one of the subjects most sus-

---

30. Bayle 1684, "Avis au lecteur," fol. 5v.
31. Bayle 1684, "Avis au lecteur," fol. 5v–6r.
32. Bernier 1684.
33. Bernier 1682.
34. Bernier 1684, 2:379–480. The main difference between the 1682 pamphlet and the "Doutes" in the second edition of the *Abrégé*, other than the inclusion of the relevant passages from Gassendi in the 1682 pamphlet, is that Bernier deletes a couple of his "Doubts" in the second edition; missing are two discussions of the void: pp. 45–48, "au lieu que nous devrions corriger nostre imagination, et concevoir que le vide n'estant rien, un corps dans le vide ne seroit en aucune chose, ou en aucun lieu"; and pp. 144–54, "si la raison qu'on apporte ordinairement pour prouver la necessité des petits vuides, est icy dans toute sa force." Both discussions are actually attempts to defend Gassendi rather than real doubts.

35. For Gassendi's theory of space, referring to the doctrines of the *Animadversiones* and *Syntagma*, see Bloch 1971, chap. 6; see also Koyré, "Gassendi et la science de son temps," and Rochot, "La vraie philosophie de Gassendi," both in *Actes du Congrès du Tricentenaire de Pierre Gassendi* 1957, 178–79, 244–45. A recent text treating Gassendi's theory of space and Bernier's critique of it is Lennon 1993, chap. 2, sec. 7; chap. 6, sec. 18.

36. For Descartes's doctrine of extension and space, see *Principles*, arts. 10–15. See also Garber 1992, chap. 5.

pect for theologians, in the debate about the Eucharist."[37] But was Bernier really being inconsistent? Did he really abandon Gassendi in his *Doutes?* Did he actually compromise the Cartesians in his *Reply to de la Ville?* Instead of simply blaming Bernier for some real or imagined slight against Descartes, one can use this episode to explicate the patterns of debates in the latter half of the seventeenth century among Cartesians, Gassendists, and Aristotelians for intellectual legitimacy in such social spheres as the court, the Church, and the university.

Let us follow Bernier's own thoughts about this issue. According to him, the problem posed by de la Ville is as follows:

> What is at stake is whether one can maintain simply with Descartes *that the essence of matter consists in extension,* or as Gassendi says, *that considering things according to the ordinary laws of nature, the essence of matter seems to consist in solidity, or impenetrability, from which extension necessarily follows.* For it is claimed that if either one or the other of these opinions is true, it follows that extension, as essential to matter, can never be without matter, nor matter without extension; this is contrary to what is commonly taught in the schools, namely, that after transubstantiation, the extension of the bread subsists without bread and the body of Christ without its extension. The essence of matter therefore does not consist in extension, nor does it consist in solidity or impenetrability, but extension must be something accidental to matter, that is, some particular accident or some small entity that makes matter be extended, and that God by his infinite power can make subsist without matter. Here in a few words are the state of the question and the foundation of the objections of M. de la Ville and of several others who have preceded him.[38]

Bernier's resolution to de la Ville's problem began with what he calls assumptions. First, he noted that church councils did not say that the real extension of the bread remained after transubstantiation, only that the body of Christ was in its own, real extension. Second, it was not the intent of the church to legitimate a particular metaphysics, to determine that the species or accidents of the bread and wine were some small distinct entities, separable from matter, and not modes of matter or some other thing. Third, the Council of Trent, referring to what remains of the bread and wine after transubstantiation, used the term *species,* not *accident.* The former signified appearances, as if the council wanted it to be understood that, after transubstantiation, by a continuation of the miracle to which Catholics must submit, the species or appearances of the bread remain, even though there

---

37. Bouillier 1858, 1:359–60.
38. Bernier, *Eclaircissement,* in Bayle 1684, 45–46; repeated as Doute 15, Bernier 1684, 2:479–80.

was no longer any bread, nor anything that could be in the bread; and that the species or appearances of the body of Christ were not in the sacrament, even though his body was truly and really there.

Given those assumptions, Bernier argued that one could respond to de la Ville by distinguishing two kinds of extension, one true and real, which is the body itself, and the other apparent, which is only the appearance of the body, or the appearance of the true and real extension. Thus, after transubstantiation, the apparent extension of the bread remained, even though the real and effective extension of the bread did not remain. And, of course, the real extension of Christ's body is in the sacrament, but his apparent extension is not. Bernier's account even used the image of the magician's trick, or of the apparition of a child in the hands of a priest officiating at mass, but with a difference: in the case of the Eucharist, although our senses are deceived, we ourselves are not deceived because we have been forewarned of the truth of the mystery.

Having sketched his resolution, Bernier then posed "a considerable difficulty" in order to turn the argument against de la Ville: "It is said that necessarily the body of Christ must be stripped of its extension and that all its parts interpenetrate each other, otherwise how can we eat it and transmit it at once into our stomach, as we do?"[39] His answer was that God, who can make a camel pass through the eye of a needle, can also bring it about that we eat Christ's flesh, actually extended, without its appearing extended to us; but it is incomparably more difficult to conceive that all the parts of a body, having no extension, interpenetrate each other, and the body remains a body: "Do you not perceive a contradiction when you say or conceive that a mountain reduced to a point would still be a mountain? In truth, M. de la Ville, it seems to me that it is very dangerous to go so fast, and that, before determining absolutely that all the parts of Christ's body are stripped of all their extension and interpenetrate each other, one must think carefully."[40]

The crux of Bernier's response was, surprisingly, very similar to that of Arnauld: the Eucharist is a mystery; God's action is miraculous; and one can allow philosophers to philosophize in their own way, as long as they propose probable, nondogmatic solutions within the sphere of the possible, limiting themselves to the natural—not supernatural—course of things:

> But without bothering with the replies of others, and without even insisting on the one I have proposed, it seems to me that de la Ville, without wounding his

---

39. Bernier, *Eclaircissement*, in Bayle 1684, 53.
40. Bernier, *Eclaircissement*, in Bayle 1684, 55–56.

conscience, could always have allowed the Gassendists to philosophize in their manner and to say, not dogmatically and decisively in the manner of Descartes, but *that considering things according to the ordinary laws of nature, the essence of matter seems to consist in solidity, or impenetrability, from which extension necessarily follows;* for this manner is completely modest; the Gassendists decide nothing positively and absolutely.[41]

Bernier's tactic was to sketch a probable resolution to the problem, limit the Gassendist's answer to the natural sphere, and then attack the dogmatists, that is, both de la Ville and Descartes.

In fact, in order to emphasize his point, the last ten pages of his *Eclaircissement* became an attack on Cartesian dogmatism: "Do you wish to know what you could have objected to Descartes? I will touch upon it a little for you, in order to amuse the reader somewhat and to make him see the wrong you do to Gassendists by not distinguishing them further from the Cartesians."[42] There followed the criticism that Descartes's doctrine that the void implied a contradiction is too far-reaching, then some polemics against the bêtes-machines, the Cartesian proof for the existence of God, and occasional causes. But what should not be lost in the final volley of criticism is that Bernier's reply to de la Ville can also be used by the Cartesians. Bernier's solution is not doctrinal; it can countenance several theories of matter and space if they are advanced cautiously. In fact, paraphrasing Bernier, a Cartesian (though perhaps not Descartes himself) can also assert that "it seems to me that de la Ville, without wounding his conscience, could always have allowed the Cartesians to philosophize in their manner and to say, not dogmatically and decisively, but that considering things according to the ordinary laws of nature, the essence of matter seems to consist in extension." And, of course, for such a Cartesian, transubstantiation in the Eucharist would be an extraordinary event; after transubstantiation, the real extension of Christ's body would be in the sacrament, but the apparent extension of the bread would remain, even though the real and effective extension of the bread would not.

Thus, it would not be inconsistent for Bernier, given his *Eclaircissement*, subsequently to criticize the Gassendist doctrine and to lean toward the Cartesian doctrine of space and place in his *Doutes*. In fact, in keeping with his nondogmatic attitude, it might even be appropriate for him to do so. And in the first of his doubts he did reject the Gassendist doctrine of an incorporeal, penetrable, and immobile space, the reference for all motion, although, in opposition to Descartes, he defended the possibility of the

---

41. Bernier, *Eclaircissement*, in Bayle 1684, 75–76.
42. Bernier, *Eclaircissement*, in Bayle 1684, 81.

void.⁴³ When reading Bernier's *Doutes,* one should keep in mind his own characterization of them in the preface to the 1682 treatise:

> These doubts are not about the foundation of this philosophy, for I do not believe that one can reasonably philosophize using a system other than that of atoms and the void; however, they are about some important matters, such as space, place, motion, time, eternity, and some others. Whether these doubts are well founded or not, you will judge. This small book will always serve you in two ways. The first is to let you see the poverty of all of our philosophies (I have been philosophizing for thirty years extremely persuaded of certain things, and yet here I begin to have doubts about them). The other, to give a rather general idea of Gassendi's philosophy, which, after all, seems to me the most reasonable, the simplest, the most sensible, and the easiest of all philosophies.⁴⁴

These thoughts were echoed in his preface to the "Doutes" from the second edition of the *Abrégé:* "I have been philosophizing for forty years extremely persuaded of certain things, and yet here I begin to have doubts about them.... However, Madam, this must not shock us, and we must not imagine that all natural things are of that degree of obscurity; philosophy, mainly the philosophy of Gassendi, always has the advantage that it allows us to discover a great number of truths which without its assistance would remain hidden to us."⁴⁵

In his *Doutes,* Bernier applied to himself the form of his own analysis with respect to the problem of the Eucharist from the *Eclaircissement.* He tried to act in the fashion of the magician who causes one to accept the appearance of criticism, while steadfastly maintaining the (Gassendist) substance underneath the appearances. Ultimately, Bernier's skirmishes with the Cartesians and the Aristotelians revealed the strategy of the Gassendists for acceptance in the intellectual climate of the latter half of the seventeenth century; by emphasizing their probabilism, their lack of dogmatism, they seem to have deflected the condemnations and criticism that the Carte-

---

43. That is, Bernier defends the possibility of intra-mundane, small voids, not of an extra-mundane void. He devotes his first three Doubts to the issues of space, place, and void: "1. Si l'espace de la maniere que monsieur Gassendi l'explique est soutenable"; "2. Si l'on peut dire que le lieu soit l'espace"; "3. Si l'on peut dire que le lieu soit immobile" ("Doutes," in Bernier 1684, 21982 405). As we have already indicated, Bernier's final doubt—"15. Si l'opinion des ancients touchant l'essence de la matiere se peut accorder avec les mysteres de la religion"—is a repeat of the problem of the Eucharist, from the *Eclaircissement,* with an added reference to chapter 4, on Qualities.

44. Bernier 1682, preface.

45. Bernier 1684, 2:379–81.

sians received and that they also could have received as defenders of a new philosophy, indeed, a philosophy that emphasizes atoms and the void.[46]

The Gassendists escaped the condemnations received by the Cartesians. However, the various condemnations of Cartesianism did not prevent Cartesianism from being discussed and even taught. The temporal authorities attempted to issue new condemnations in 1704–5, and the Jesuits formally condemned thirty Cartesian propositions in 1706, including some concerning the Eucharist and one against the law of inertia. Among the prohibited propositions were: "There are no substantial forms of bodies in matter," "There are no absolute accidents," and "There is, in the world, a precise and limited quantity of motion, which has never been augmented nor diminished."[47] As Arnauld predicted, the latter condemnations also failed to succeed.

Either the authorities in France were not ruthless enough to have their will carried out or their will was not unified enough to give efficient orders. Of course, we all know that Bruno was burned at the stake in 1600, Campanella imprisoned and tortured from 1599 to 1626, and Galileo condemned in 1633 and put under house arrest from then on, but those events occurred under the sphere of influence of Rome. In France, Vanini was hanged and burned at the stake in 1619. But the atomists Villon, de Clave, and Bitaud were simply banished from Paris. De Clave continued to publish chemical works well after 1624. Again, given the division of powers in France, formal condemnations were not very effective. Some of the more daring thinkers found protectors such as Cardinals Bérulle and Richelieu, whose political agendas did not coincide with those of Rome. Without more effective repression, Cartesianism could not be halted in France.

---

46. Of course, this is a single study, based on a few thinkers and a few texts. It depends upon the extent to which thinkers such as Arnauld and Bernier and their positions can be thought as typical.

47. "Prohibited propositions by Michel-Angelo Tamburini, General of the order in 1706," in Ariew 1994a; trans. in ACS 258–60.

[10]

Scholastic Critics of Descartes: The *Cogito*

In contributing to the massive twentieth-century writings on Descartes's *cogito*, my intent is not to ask more logico-linguistic questions about it but rather to investigate the criticisms it received from seventeenth-century philosophers, in part for what they can tell us about these philosophers and in part for what they can reveal about Descartes and the *cogito* itself. First, however, I wish to place the criticism in its proper context. I should emphasize that critiques of the *cogito* are relatively rare, though not nonexistent, and that attending to them solely, to the exclusion of the other interests of seventeenth-century philosophers, might be misleading.

For twentieth-century philosophers, Descartes is the person who redirected philosophy inward with the *cogito*. He is the father of modern philosophy, a thinker whose primary motivations were epistemological, in opposition to the metaphysico-theological concerns of the Scholastics; he is said to have been obsessed by the establishment of secure foundations for knowledge or the search for a new method for the acquisition of knowledge. Following Descartes, with the *cogito* as the turning point, moderns no longer needed to ask scholastic questions about being, but progressed instead to reflective questions about the self (bringing themselves a step closer to the linguistic turn).[1] However, that image of Descartes does not

---

1. In a recent interesting discussion in the *British Society for the History of Philosophy Newsletter* (n.s., 1. no. 1 [1996]: 29–33), "a small but eminent group of Descartes scholars" were asked to address briefly the questions "Why should we teach Descartes to philosophy students?" and "Are there any aspects of his philosophy which are still living?" Four out of five (Vere Chappell, Stephen Gaukroger, Tad Schmaltz, and Steve Wagner) responded by making reference to Descartes and epistemology: "his foundationalist program in epistemology," "the royal road to problems of epistemology," Descartes as "the pure epistemologist," and the *Meditations* as offering "a distinctive form of epistemic review." The fifth (Richard Watson) talked about mind-body dualism.

mesh very well with the reality of the reception of his philosophy in the seventeenth century. Rarely was Descartes the epistemologist discussed then; instead, and not unexpectedly, I suppose, seventeenth-century critiques were predominantly directed against Descartes's various metaphysical theses. The history of condemnations of Cartesianism in the seventeenth century provides a vast amount of evidence for this different image of Descartes. Seventeenth-century civic, religious, and scholastic authorities were most often extremely unhappy with Descartes the revisionist metaphysician, with such doctrines as the denial of substantial forms, the principle that extension is the essential attribute of matter, and the claim that the universe is singular and indefinite.[2] In fact, the controversy over these metaphysical doctrines and their consequences almost eclipsed all other discussions of Cartesianism.

One can read whole books critical of Descartes's philosophy, written in the seventeenth century without running into any discussion of the *cogito* or any other aspect of Descartes's epistemology.[3] Witness, for example, Jean-Baptiste de la Grange's two-volume work, *Les principes de la philosophie contre les nouveaux philosophes, Descartes, Rohault, Régius, Gassendi, le p. Maignan, etc.*[4] De la Grange, a French Oratorian, believed Descartes to be a dangerous person, a thinker whose philosophy had rightly been condemned by the king, since it was based on principles that were inconsistent with Christian theology. He also thought that Descartes's philosophy was ruining Christian theology by undermining the scholastic philosophy upon which it had been based:

> Although one might have an inclination for Descartes's philosophy because it appears new and much easier than that of the Peripatetics, nevertheless, were one to know its principles just a little, one would easily see that this doctrine contains something bad, such that it should be surprising that so many intellectuals profess the doctrine. For it is not necessary to enter deeply into the details of the propositions taught by Descartes to know that it is for good reason that his majesty . . . has not long ago prohibited the opinions of that author from being taught in his kingdom. It suffices to know that these principles ruin a large part of theology by completely destroying ordinary philosophy. . . . One need only hear Descartes explain the great mysteries of the faith in a completely new way, and claim that all Catholic theologians up to now were mistaken, to be persuaded that whether his doctrine is erroneous or not, at least it is dangerous, and that the professors of philosophy are wholly wrong in

---

2. These principles were condemned at Louvain. See Chapters 8 and 9.
3. The same can be said about works generally sympathetic to Descartes's philosophy: Bossu 1674, for example.
4. Grange 1682.

teaching it to young people, in whom it is not good to inspire the love of novelty nor the hatred for the ancient doctrine.⁵

In consequence, de la Grange discussed Descartes's principles most critically, beginning with the rejection of the plurality of worlds.⁶ According to de la Grange, Descartes's thesis was based on the definition of matter as extension, the indefinite extension of the world, and the assumed erroneous principle that two bodies cannot occupy the same place:

> What I find amusing is that Descartes boldly teaches extremely dangerous conclusions that he derives from two completely unproven principles. The first is that he assumes that everywhere there is space there is also matter. . . . The principle that he must necessarily assume, in order to conclude that several worlds are impossible—which nevertheless he does not mention—is that two bodies cannot, absolutely speaking, be in the same place, and that matter cannot be in some other matter. . . . Thus one should note that not only is Descartes's conclusion, that several worlds are impossible, false and dangerous, but also that it is derived from a dangerous principle, that two bodies cannot be, absolutely speaking, in the same space.⁷

De la Grange continued, in succession, with such topics as whether animals can reason, the accidents of the Eucharist, the nature of place and the void, and the infinity of the world; in his second volume, he broached the topic of the immobility of the earth and other similar subjects. His primary motivation was the reestablishment of the Scholastics' substantial forms; he did not seem at all interested in skepticism, hyperbolic doubt, the *cogito*, ideas, or the analysis of sense perception.

A visit to Descartes's world, guided by the Jesuit Gabriel Daniel, imparts the same image.⁸ Daniel satirized the Cartesian doctrines he found most offensive, namely, the union of soul and body (Descartes separating and reuniting them when he pleased), the account of motion and the conservation of quantity of motion, the explanation of the Eucharist, the denial of the void, the acceptance of vortices and the motion of the earth, and the irrationality of animals—that is, Descartes's bêtes-machines. Again, the topics discussed in his work, which was extremely popular,⁹ related to metaphysics, theology, physics, and cosmology but not to epistemology.

5. Grange 1682, 1–3.
6. Grange 1682, 6–7.
7. Grange 1682, 7–9.
8. Daniel 1690, 1693.
9. The work was even translated into Latin and English, the latter as Daniel 1692.

This is not to say, however, that all seventeenth-century critiques were exclusively about metaphysical issues. Now and then one can find bits and pieces of the topics that might speak to more modern concerns. Although the Jesuits of the Collège de Clermont largely rejected Cartesianism in 1665, aiming most of their criticism at the usual metaphysico-theological suspects, one can also glimpse a criticism—under the rubric of "what is distasteful to mathematics"—of what Scholastics called the classification or subalternation of the sciences, that is, the set of doctrines discussed in conjunction with Aristotle's *Posterior Analytics*.[10] The basic issue was whether mathematics should be subalternated to physics, as most Scholastics thought, or whether that order needs to be overturned, as seems to be Descartes's doctrine—in other words, whether or not mathematics, as an abstraction from natural things (that is, from the objects of physics), can be used in the explanation of natural things. In addition, the first criticism, under the rubric of "what is distasteful to philosophy," can be interpreted as a vague pragmatic critique of Cartesian doubt.[11]

---

10. Reported by Oldenburg in a letter to Boyle: "The Cartesian hypothesis must be distasteful to mathematics . . . because it is applied to the explanation of natural things, which are of another kind, not without great disturbance of order" (Oldenburg 1966, 2:35). See Chapter 8 for the full text. The general issue is based on the principle that, among other requirements, genuine scientific knowledge needs to be knowledge of the reasoned fact (or the reason why), not of the mere fact (Aristotle, *Posterior Analytics* I, chap. 13); thus demonstration is a syllogism that proves the cause—i.e., the reasoned fact, not the fact (Aristotle, *Posterior Analytics* I, chap. 24). But reasoned fact and fact differ both when they are investigated by a single science and by different sciences. Aristotle can give examples of geometers knowing the reason why, when natural philosophers know only the fact; this situation occurs when fact and the reasoned fact are investigated by different sciences, that is, when problems are related to one another as subordinate and superior, as in the case of the mathematical sciences (such as optics or astronomy) and mathematics. The question, then, involves the ultimate classification of the sciences, whether mathematics is subordinate to natural philosophy or vice versa. Aristotle's doctrine is complex and open to interpretation on such topics, but Thomistic interpretations of Aristotle—what the Jesuits were generally committed to—are more rigid about such matters. Thomas holds that mathematicians abstract from sensible matter and motion (*Commentary on the Metaphysics* V, lect. 16, n. 989, and elsewhere) and that the mathematical sciences prove the same conclusions as the naturalists by formally different principles of demonstration (*Summa Theologiae* IIa.IIae, q. 1, art. 1). This is consistent with Thomas's discussion of the subalternate sciences: in the mathematical sciences, the geometer explains the reason why according to the formal cause, but the quantitative form is a remote cause as far as the natural phenomenon is concerned (*Commentary on the Posterior Analytics* I, chap. 13). For Thomas, mathematics and the mathematical sciences are subalternated to natural philosophy. Mathematics looks to natural philosophy for its justification.

Even the great Jesuit mathematician Christopher Clavius, accepts these doctrines, though he tries to mitigate the implicit criticism in them. He limits himself to prudential considerations when discussing the subalternation of the sciences. For more on this issue, see Ariew 1992.

11. "The Cartesian hypothesis must be distasteful to philosophy . . . because it overthrows all its principles and ideas which commonsense has accepted for centuries" (Oldenburg 1966, 2:35).

Some of the condemnations of Cartesianism listed propositions that cannot be best described as evincing metaphysico-theological concerns. Statements condemning Cartesian doubt, for example, are among the propositions censured by the University of Paris in 1691 or condemned by general of the Jesuits in 1706.[12] Moreover, when the Oratorian professors at Angers were expelled from their positions in the 1670s for teaching Cartesian philosophy, among the reasons given for the expulsions was that they did not reject Descartes's skeptical method. As usual, they were condemned for having denied real accidents and the finiteness of the world; but they were also reproached for having forgotten that "to say that one must doubt all things is a principle that tends toward atheism ... or at least toward the heresy of the Manicheans."[13] And they were censured for having accepted Descartes's *cogito* as a principle of knowledge. For example, one of the mistakes made by a certain Oratorian professor was that

> he does not perceive that it is impossible for the reasoning to be the first principle of reasoning and knowledge, otherwise a thing would be a principle of itself. Now, *cogito ergo sum* is an argument, but a truly defective one, since the consequence of this enthymeme is the same thing as the antecedent. For *cogito* means in philosophical terms *Ego sum cogitans*. . . . The first principle of the sciences must be universal and necessary, because science is of *universalibus et necessariis;* and this principle *cogito* is something singular and extremely uncertain, since it is . . . (as the philosophers say) *de individuo et in materia contingente. Ego sum: est propositio singularissima, et, cogito est quid incertissimum;* and is there a student who does not know that the first principle is the foundation of the truth of demonstrations and was never an *a posteriori* demonstration by effects? Now, *cogito ergo sum* proves *a posteriori* the existence of man by means of his own operation.[14]

The authors of the document reject the *cogito* as a principle of knowledge, or as science, properly speaking, because such a principle, according to the *Posterior Analytics,* must be a "commensurate universal," a proposition whose predicate belongs essentially to every instance of its subject (73b 26–30). The *cogito,* thus, does not fit the scholastic model for pure scientific knowledge at all. It is neither universal nor necessary, but singular and contingent. Moreover, it is an argument, even a defective one; either it is dependent upon an unspecified major premise or it begs the question. And if it is an argument, it cannot be a principle of knowledge: an argument cannot itself be a principle. This compact passage relates succinctly

---

12. Argentré 1736, pt. 1, p. 149; Rochemonteix 1899, 4:89n–90n.
13. Babin 1679, 41.
14. Babin 1679, 42.

many of the reasons why seventeenth-century philosophers rejected the *cogito*; indeed, it is a paradigm of their criticisms of it.

I shall return to this point, but first I wish to recall briefly the critiques of the *cogito* that seventeenth-century philosophers would have been familiar with, given that they were published as objections to the *Meditations* in Descartes's single volume of *Meditations, Objections and Replies*. These texts include the famous objections by Hobbes and Gassendi, respectively in the *Third* and *Fifth Objections*, and also some lesser-known ones by the anonymous objectors of the *Sixth Objections* and by the Jesuit Pierre Bourdin in the *Seventh Objections*, published only with the second edition of the *Meditations*.[15] Hobbes's objection is directed against the *res cogitans*, as opposed to the *cogito* itself. He allows that "I exist" follows from the fact that "I think," but he objects that "when the author adds 'that is, I am a mind, or intelligence, or intellect or reason,' a doubt arises."[16] Hobbes is objecting to the argument in which Descartes reasons from "I am thinking" to "I am a thought"; according to Hobbes, Descartes might just as well have said "I am walking, therefore I am a walk." Descartes sees the objection as a threat to the *cogito* itself, because he himself brings it up in discussion of Gassendi's dismissal of the *cogito*. Gassendi objects: "You conclude that this proposition, *I am, I exist*, is true whenever it is put forward by you or conceived in your mind. But I do not see that you needed all this apparatus, when on other grounds you were certain, and it was true, that you existed. You could have made the same inference from any one of your other actions, since it is known by the natural light that whatever acts exists."[17] In his reply to Gassendi, Descartes denies that one can infer one's existence from any of one's actions, and in particular, he denies that "I am walking, therefore I exist" would be known with as great a certainty as "I am thinking therefore I exist," or with as great a certainty as "I think that I am walking, therefore I exist." It would seem that for Descartes there is a difference in the degree of certainty one can have about one's thinking as opposed to one's walking.[18] Descartes thus safeguards one of the more important functions of the *cogito*, the conclusion that the mind is better known than the body. As a materialist, Hobbes is committed to rejecting that conclusion and thus seeks to undermine the *cogito* by claiming that one's existence can be inferred

---

15. It should be pointed out that sets of objections and replies were circulated with the *Meditations*, so that later objectors replied to the *Meditations* and to previous sets of objections and replies.
16. AT 7:172; CSM 2:122.
17. AT 7:259–60.
18. That is what Gassendi appears to deny in his reply to Descartes's reply (Gassendi 1962, 82).

from any of one's acts.[19] His general strategy is to deny that the I who is thinking is anything other than a body. However well received it would be by later thinkers, this move has little appeal for the nonmaterialist majority position.

The anonymous Sixth Objectors announce another, more promising line of attack, though they embed it in an obscure argument within a series of peculiar arguments: "from the fact that we are thinking it does not seem entirely certain that we exist. For in order to be certain that you are thinking you must know what thought is or thinking is, and what your existence is; but since you do not yet know what these things are, how can you know that you are thinking and that you exist? Thus neither when you say 'I am thinking' nor when you add 'therefore I exist' do you really know what you are saying."[20] The objectors go on to assert that knowing that one is saying or thinking anything requires one to know that one knows what one is saying, which requires one to be aware of knowing that one knows what one is saying, and so forth. Descartes does not find it at all difficult to deny that the *cogito* requires reflective knowledge, demonstrative knowledge, or knowledge of reflective knowledge. He also easily dismisses a couple of other, even more peculiar, objections by the Sixth Objectors. However, the main objection appears to survive. Gassendi gives a precise exposition of it in his *Disquisitio Metaphysica* and brings out a side issue: The *cogito* is an enthymeme that requires the major premise "he who thinks exists," and thus the *cogito* cannot be the first truth discovered.[21] Descartes seems to accept part of the criticism when he admits, in his reply to the Sixth Objectors, that "it is true that no one can be certain that he is thinking or that he exists unless he knows what thought is and what existence is."[22]

In the *Seventh Objections,* Bourdin tries to expand the list of what one needs to know in order to conclude that one exists, though he intentionally expresses himself in cryptic fashion. Not only do we need to know what

---

19. In this respect Gassendi's critique can be lumped in with that of Hobbes. Of course, Gassendi is not a materialist, but this is not at all evident in his critique of the *cogito* or in the *Fifth Set of Objections* as a whole—witness his calling Descartes "O Mind" (*ô Mens*, AT 7:275 and elsewhere) and Descartes's replying "O Flesh" (*ô Caro*, AT 7:359 and elsewhere).
20. AT 7:413; CSM 2:278.
21. Gassendi 1962, in Meditationem II, Dubitatio I, art. 6, p. 84. See also Goudin 1864, vol. 4, quest. 1, art 1, p. 282: "Mais Descartes est absolument intolérable, lorsqu'il veut que le sage renonce, au moins pour un temps, à tous les *principes,* parce qu'ils sont tous douteux, et quand il nous engage à commencer la connaissance des choses par ce *principe: Je pense,* pour en inférer aussitôt: *donc je suis;* car, pour n'en pas dire plus long, si l'âme écarte comme douteux notre *premier principe* avec tous les autres, elle devra douter aussi *ce qui pense est ou n'est pas.* Il deviendra possible de penser et de n'être pas, s'il est possible que la même chose soit et ne soit pas. Ce *principe* de Descartes ou plutot cet enthymème s'appuise donc sur notre *principe.*"
22. AT 7:422; CSM 2:285.

thought is, but also that dreaming entails thinking and not vice versa, since dreaming is not identical to thinking:

> I am dreaming that I am thinking. So I am not thinking.
> "No," you reply, "if someone is dreaming, then he is thinking."
> I see a ray of light. Dreaming is thinking, and thinking is dreaming.
> "Certainly not," you say. "Thinking extends more widely than dreaming. He who is dreaming is thinking; but he who is thinking is not dreaming all the time, but may be thinking while awake."
> But is this right? Are you dreaming it, or are you really thinking it? If you are dreaming that thinking extends more widely, does it follow that it really does so? ... how do you know that thinking extends more widely than dreaming?[23]

Bourdin wishes to show that Descartes needs to presuppose much knowledge in order to be able to conclude that he exists. But this is not the main thrust of his criticism. He understands that Descartes claims not to need syllogisms: "Your method denigrates the traditional forms of argument, and instead grows pale with a new terror. ... If you propose any syllogism, it will be scared of the major premise, whatever it may be. 'The evil demon may be deceiving us,' it says. What about the minor premise? It will tremble and call the minor premise doubtful. ... Finally, what about the conclusion? It will run away from all conclusions as if they were traps and snares."[24] So Bourdin's objection is actually that without syllogism, without the ability to go from truth to truth, and without a first truth, Descartes can never break out of what seems to him to be so into what really is so:

> "I am thinking," you say.
> I deny it; you are dreaming that you are thinking. ...
> "I exist, as long as I am thinking," you say. ... "This is certain and evident," you continue.
> No; you merely dream that it is certain and evident.[25]

Bourdin's argument is ridiculed by Descartes: "These comments are amusing enough, if only because they would be so inappropriate if they were intended to be serious." He merely reasserts that he would like people to remember that "if something is clearly perceived, then no matter who the perceiver is, it is true, and does not merely seem or appear to be true."[26]

---

23. AT 7:494; CSM 2:334.
24. AT 7:528; CSM 2:359 (slightly altered).
25. AT 7:498; CSM 2:337.
26. AT 7:511; CSM 2:348. See also AT 7:461–62 (CSM 310), where Descartes makes the same claim.

Descartes and Bourdin are at an impasse. Bourdin thinks that madmen or dreamers never get to the truth, only to what seems true to them; since they lack traditional methods of reasoning, they cannot arrive at a first truth or any other truth.

Let us return to the question of whether the *cogito* is a syllogism that requires the major premise "everything that thinks exists." Descartes already foresaw this objection and attempted to defend himself against it in his replies to the *Second Objections*. The Second Objectors assert that, according to Descartes's own words, the *cogito* requires knowledge of the existence of God, something he has not achieved by Meditation II.[27] Descartes, defending himself, replies that "awareness of first principles is not called knowledge" and that "when we become aware that we are thinking things, this is a primary notion which is not derived by means of any syllogism." He adds, "When someone says 'I am thinking, therefore I am, or I exist,' he does not deduce his existence from thought by means of a syllogism, but recognizes it as something self-evident by a simple intuition of the mind. This is clear from the fact that if he were deducing it by means of a syllogism, he would have to have had previous knowledge of the major premise 'Everything which thinks is, or exists.' "[28] According to Descartes, then, the *cogito* is therefore not knowledge, properly speaking, but a simple intuition; it is not a syllogism. Does it depend upon knowing what thought is, what existence is, etc., as Descartes appears to say in *Replies VI*? Or is it independent of such knowledge, including the major premise "everything which thinks is, or exists," as he seems to say in *Replies II*? Descartes appears to accept the former when he claims in *Principles* I, article 10: "When I said that the proposition *I am thinking, therefore I exist* is the first and most certain of all to occur to anyone who philosophizes in an orderly way, I did not in saying that deny that one must first know what thought, existence and certainty are, and that it is impossible that that which think should not exist, and so forth. But because these are very simple notions, and ones which on their own provide us with no knowledge of anything that exists, I did not think that they needed to be listed."[29]

In fact, one of the questions addressed to Descartes by Franz Burman was whether his assertion in *Replies II* is contradictory with what he affirms in *Principles* I, art. 10.[30] Descartes's reported answer is interesting: He allegedly distinguished between what he is expressly and explicitly aware of before the *cogito* and what is implicitly presupposed by it. He is said to have

---

27. AT 7:124–25; CSM 2:89.
28. AT 7:140; CSM 2:100.
29. AT 8A:8; CSM 1:96.
30. AT 5:147; CSM 3:333.

claimed, using the language of syllogism, that "before this conclusion, 'I think therefore I am,' the major 'whatever thinks is' can be known; for it is in reality prior to my conclusion and my conclusion depends on it."[31] Descartes's answer to Burman was not known during the seventeenth century. If it had been, it would have provided plenty of ammunition for his critics; regardless, it would seem that the way is clear to object to the *cogito* that it is some kind of argument (if not a syllogism) which depends upon principles that are not proven or subject to doubt. One can also attempt to undermine the claim that it is a simple intuition.

Both of these strategies are used in the most extended critique of the *cogito* by Pierre-Daniel Huet, then bishop of Avranche, an important figure in late-seventeenth-century French intellectual circles.[32] Huet repeats the objection that the *cogito* is an enthymeme missing its major premise or that it begs the question. He asserts that Descartes has abandoned his promise to take everything as doubtful when he accepts as certain something that is doubtful and should be held to be false.[33] He adds to the list of what needs to be known in order for "I exist" to follow from "I think." He reminds his readers that there are many propositions that precede the *cogito,* such as "everything that thinks exists" and even "everything that acts exists." He then argues that "we cannot know that that which acts exists unless we know what it is to act and what it is to exist. And to know what it is to act, we must know what is the agent, what is its cause, how it acts, and why it acts. Moreover, to know what it is to exist, we must know what the thing is that exists, what the cause is for it to exist, how it exists, and why it exists."[34] He further argues that Descartes cannot defend himself by means of the rules of logic, since he has resolved to hold all things as false and cannot have any faith in them. He adds a new element to these kinds of objections: since, according to Descartes, God can make two contradictory propositions be true at the same time, God can make it be that he who thinks both exists and does not exist at the same time, or that he who thinks does not exist. The *cogito* cannot be an absolute certainty.[35]

In his lengthiest objection, he examines the temporal nature of the *cogito*. He argues, in a number of different ways, that the *cogito* occurs in time,

---

31. AT 5:147; CSM 2:333 (slightly altered).
32. The critique is contained in chap. 1, art. 5–13, pp. 21–38, of Huet 1689. It is interesting to note that the subtitle of the work is *Servant d'éclaircissement à toutes les parties de la philosophie, sur tout a la métaphysique*. On the importance of Huet to French society, see Lux 1989. See also Malbreil 1991.
33. Huet 1689, chap. 1, art. 5; see also art. 7.
34. Huet 1689, chap. 1, art 7.
35. Huet 1689, chap. 1, art. 6; see also art. 13. Presumably, this element appears principally because of the publication of Descartes's correspondence in 1657–67.

that it requires the faculty of memory, and that it fails on account of it. Basically, the argument is that the *cogito* cannot be expressed as "I think, therefore I am." "I think" and "I am" can never occur at the same moment, so that the *cogito* can only be "I think, therefore I was," "I think, therefore I will be," "I thought, therefore I am," "I thought, therefore I will be," or "I thought, therefore I was." And, of course, none of these are satisfactory. The simplest argument is that the *cogito* requires that everything that thinks exists during the time it thinks. "But my thought has stopped existing when I say 'therefore I exist' and the time in which I say 'I think' is different than the time in which I say, 'therefore I exist,' which is why the argument signifies 'I think, therefore I will be,' or 'I thought, therefore I am.' "[36] Huet also argues that when Descartes says, "I think," the object of his thought is his thought itself. But since a thought cannot both be an act and the end toward which the act is referred, Descartes's thought as object is not the thought by which his mind thinks: the thought by which Descartes thinks is different than the one about which Descartes thinks. "I think" is then "I think that I think," which actually signifies "I think that I thought." And, of course, Cartesians cannot conclude "therefore I exist" from that.[37]

Huet also denies that the *cogito* can be a simple intuition. According to him, "If 'I think therefore I exist' were a simple action of the mind, it would not be true that 'I think' would be better known than 'I exist.' "[38] As evidence for the claim that "I think" is better known than "I exist," Huet affirms that you can deduce "I exist" from "I think," but you cannot deduce "I think" from "I exist." Thus, "I think" must be known prior to "I exist" and "I think, therefore I exist" cannot be a simple intuition or action of the mind, but a progression of knowledge, the acquisition of something unknown from something known—in other words, an argument.

Pierre-Sylvain Régis,[39] acting as Descartes's stand-in—Descartes having been dead for four decades—has no difficulty in replying to Huet's critique. He denies that the *cogito* is an enthymeme or that it begs the question. He asserts that Descartes has not abandoned his promise to doubt everything when he accepts something as true after having examined it.[40] He claims that Descartes never accepted the general rule to hold everything as false, but merely resolved to consider as false whatever appears doubtful. Régis distinguishes between real doubt, arising from the nature of things, and a feigned, methodological doubt—what Descartes called hy-

---

36. Huet 1689, chap. 1, art. 9.
37. Huet 1689, chap. 1, art. 9.
38. Huet 1689, chap. 1, art. 11.
39. Author of a textbook exposition of Cartesian philosophy (Régis 1690), among other works.
40. Régis 1691, I, art. 5.

pothetical, hyperbolic, and metaphysical doubt—arising from his resolution to doubt.[41] In keeping with this interpretation of Descartes, he asserts that Descartes held the rules of logic as false only "hypothetically" in order to examine them. He asks rhetorically: "who can prevent Descartes from holding them as true, if they have appeared to him as such, after he has examined them?"[42] On the question of whether God can make something exist and not exist or something think and not exist, Régis allows that Descartes said something like that somewhere, but he claims that Descartes said it with respect to God's extraordinary power, not with respect to things considered according to the ordinary course of nature, which is what is at stake in the *cogito*.[43] Régis also denies that "I think, therefore I exist" requires the major premise "Everything that thinks exists." We know singular propositions before general ones.[44] According to Régis, "Everything that thinks exists" does not precede "I think, therefore I exist" for those who seek to discover their existence by analysis; it only precedes the latter in the minds of those who wish to prove their existence to others by synthesis.[45] As for the argument that we cannot know we exist unless we know what it is to act and to exist and what is the agent, its cause, how it acts, why it acts, etc., he agrees that that would be required for *adequate* knowledge, but that what is at stake in the *cogito* is the *simple* knowledge of one's own existence.[46]

On the lengthy question about the temporality of the *cogito*, Régis claims that memory might be defective at times, for example, when we are dealing with distant memories (which is not the case here). However, it can be trusted at other times, when we are attending carefully to the matter at hand, as with the *cogito*. Regardless of that problem, he claims that "I think" and "therefore I exist" are both in the mind at the same time: one of them is in the understanding and the other in the will: "For one has to note that Descartes teaches expressly that thoughts are passions that belong to the understanding and that affirmations and negations are actions that belong

---

41. Régis 1691, I, art. 1.
42. Régis 1691, I, art. 6.
43. Régis 1691, I, art. 6.
44. We should add that that is Descartes's reported view in the *Conversation with Burman;* see AT5:146–47; CSM 3:332–33.
45. Régis 1691, I, art. 7. It is interesting to note that Leibniz shares this view (taking analysis and synthesis to correspond to order of knowledge and order of nature, respectively). Leibniz, of course, is almost never interested in the *cogito*. But, in the very early "Letter to Foucher" (1676), he asserts: "But even though the existence of necessities is the first of all truths in and of itself and in the order of nature, I agree that it is not first in the order of our knowledge. For you see, in order to prove their existence I took it for granted that we think and that we have sensations. Thus there are two absolute general truths, that is, two absolute general truths which speak of the actual existence of things: the first, that we think, and the second, that there is a great variety in our thoughts. From the former it follows that we exist" (Leibniz 1989, 2).
46. Régis 1691, I, art. 7.

to the will."⁴⁷ Régis easily disposes of the other objections that make the *cogito* temporal. He denies that "I think" is equivalent to "I think that I think" and, thus, he does not consider whether "I think that I think" actually signifies "I think that I thought." According to him, the *cogito* is also not temporal because there is only one thought in it. While thinking, the meditator perceives that he thinks by a single and simple thought, which is known by itself; otherwise there would be an infinite progression of thoughts.⁴⁸ Finally, he reaffirms that the *cogito* is a simple intuition and he rejects the relevance of the argument that one can derive "therefore I exist" from "I think" but not vice versa: "Since being is something more general than thought, . . . one can truly conclude that something exists from the fact that it thinks, but one cannot infer in the same way that something thinks from the fact that it is. This is sufficient to destroy the reasoning of the author of the *Censura*."⁴⁹

The matter did not end there. Régis's reply to Huet was analyzed by Jean Duhamel, professor emeritus at the University of Paris,⁵⁰ who published *Reflexions critiques sur le système cartésien de la philosophie de mr. Régis*.⁵¹ Duhamel devotes two of his chapters to the *cogito*. After considering Huet's and Régis's arguments, Duhamel comes down squarely on Huet's side. The *cogito* begs the question. Moreover, it is a defective argument.⁵² Duhamel argues that one cannot separate knowledge of one's existence from other knowledge: "The analysis of our author . . . assumes that I can separate knowledge of the heavens, of the earth, or of the sea, from its proper object; now, it is false that I can separate this knowledge from its object, any more than from its subject, for knowledge is not less relative essentially to its object than to its subject, a difference being no less essential than a genus."⁵³ He also argues that since, according to the Cartesians, God can make it be that I think and not exist, "I exist" does not follow necessarily from "I think."⁵⁴

---

47. Régis 1691, I, art. 9.
48. Régis 1691, I, art. 9.
49. Régis 1691, I, art. 11.
50. Author of a highly regarded (posthumously published) textbook (Duhamel 1705).
51. Duhamel 1692.
52. Duhamel 1692, chap. 13.
53. Duhamel 1692, chap. 13.
54. Here is Duhamel's complete argument:

De plus, on soutient que, de ce que je pense, il ne s'ensuit pas nécessairement que j'existe dans le principe des Cartésiens; car si Dieu peut faire que je pense et que je n'existe pas; de ce que je pense, il ne s'ensuit pas nécessairement, que j'existe: or Dieu peut faire que je pense et que je n'existe pas dans le principe des Cartésiens, et sur-tout de notre Auteur, qui dit expressément: *il reste donc qu'il n'y ait point d'impossibilité avant le décret de Dieu, en telle sorte que quand je dis, qu'il est impossible qu'une chose soit et ne soit pas, cela ne signifie autre chose, si ce n'est que Dieu a voulu qu'une chose fust tandis qu'elle seroit* [livre 1, part 1, de sa métaphysique, chap. 13]; ce qui prouve sans commentaire, que si Dieu vouloit

But Duhamel's greatest interest is in denying that the proposition "I think, therefore I am" is the first of all propositions known. He rejects Régis's claim that singular propositions are known before general ones and asserts that there are several general propositions that may be known before "I think therefore I exist," namely, "Everything that exists exists necessarily while it exists," "Everything that acts exists," and "Everything that thinks exists."[55] He has a number of reasons for this. Direct propositions are known before reflexive propositions, because the object precedes the knowledge or idea of which it is the object.[56] The general propositions above are direct since they refer to an external object, and the *cogito* is reflexive since it has thought and existence as its object.[57] In addition, it cannot be the case that general propositions presuppose all the particular propositions that could be subsumed under them and, in particular, the

---

par une volonté éternelle, ainsi qu'il veut autres choses qu'il veut, il seroit possible qu'une chose fut et ne fut pas; à plus forte raison, qu'elle pensast et qu'elle n'existast pas. Donc de ce que je pense, il ne s'ensuit pas nécessairement, dans les principes des Cartésiens, que j'existe. (Duhamel 1692, chap. 13)

55. "On soutient au contraire, qu'il y a plusieurs propositions qui peuvent être connues avant celle-cy, et notamment que ces propositions générales, *Tout ce qui existe, existe nécessairement pendant qu'il existe: Tout ce qui agit, existe: Tout ce qui pense, existe,* peuvent être connues avant elle" (Duhamel 1692, chap. 14).

56. "Parce que les propositions directes sont connues avant les réflexes, ce qui est connu directement, est plutost connu, que ce qui est connu par réflexion; car l'objet précède la connoissance ou l'idée dont il est l'objet, et la connoissance directe est l'objet de la réflexe" (Duhamel 1692, chap. 14).

We should point out that this is not something made up just to attempt to defeat the cogito, but a standard bit of scholastic philosophy. For example, one can read in Eustachius's 1609 *Summa philosophiae quadripartita*:

> The intellect knows other things before it knows itself. For direct knowledge is prior to reflexive knowledge, and the intellect knows things other than itself by direct cognition, while it knows itself only by reflexive cognition.... The intellect knows material substances before it knows immaterial and spiritual ones ... The intellect has prior knowledge of composite substances than of their parts or differences. For confused cognition comes before distinct cognition, and composite substances are first of all known by a confused kind of cognition, while their parts are known only by distinct cognition. Accidents are known prior to substances. For accident are generally accessible to the senses, but substances are hidden and not sensible in themselves; hence they are not so swiftly or easily known. (Physica, pars 3, tract. 4, disp. 2, quaest. 6)

It is interesting to compare the above to the post-Cartesian pronouncements of Leibniz in *De Veritatibus, de Mente, de Deo, de Universo:* "I admit that the proposition 'I think' must occur first in the order of philosophising; that is, if the primary truths are arranged in order, it will be first. For it is simpler to start from one subject of a primary proposition of experience than from its various predicates" (Leibniz 1992, 57).

57. "Les propositions générales, cy devant rapportées, sont directes, puisqu'elles tendent à un objet extérieur, et qui est en dehors de nous; car l'existence, l'action, et la pensée des autres de nous, est un objet extérieur; au contraire cette proposition, *je pense, donc je suis,* est réflexe, puisqu'elle a pour objet la pensée et l'existence, qui est en nous mêmes, et partant les propositions générales cy devant rapportées, précèdent cette proposition particulière, *je pense, donc je suis*" (Duhamel 1692, chap. 14).

general propositions above can be known before the *cogito*.[58] But Duhamel's main point is that a single item of knowledge cannot be essentially knowledge of itself, just as a single action or passion cannot act on itself or receive itself, and a single item of knowledge cannot have two formal objects.[59] He concludes:

> From which it follows evidently 1. that thinking, as it is understood by the Cartesians in this proposition, "I think," is not represented essentially to the understanding without a different perception. 2. I do not perceive immediately the ideas in me, but objects external or internal to my mind, according to whether my knowledge is direct or reflexive. From which it follows, finally, that if by consciousness the Cartesians understand knowledge of their knowledge, consciousness is knowledge distinct from the first [that is, knowledge] and can no more be accomplished without an idea than the first.[60]

Régis replied selectively to Duhamel; he denied the charge that the *cogito* is a petition of principle[61] and that one cannot separate knowledge of one's existence from other knowledge.[62] Again he rejected the argument that the

---

58. "Parce que les propositions générales supposent à la vérité quelques propositions singulières; mais il est certain qu'elles ne supposent pas toutes les propositions singulières: autrement les propositions générales exigeroient l'induction de toutes les particulières sans exception, ce qui est reconnu pour faux en matière nécessaire, dont il s'agit" (Duhamel 1692, chap. 14).

59. Here is Duhamel's full argument:

> Parce que celuy qui connoit les autres propositions connoissoit nécessairement et essentiellement connoissance d'elle-même: or il est impossible que la même conoissance soit essentiellement conoissance d'elle-même, car en ce cas l'action agiroit sur elle-même, ou la passion recevroit d'elle-même, puisque la conoissance est une action ou une passion: or il est impossible qu'une même action ou passion indivisiblement agisse sur elle-même, ou reçoive d'elle-même; et partant la conoissance, à moins qu'elle ne soit infinie dans le genre de conoissance ne peut être conoissance, d'elle-même.
>
> Une même conoissance ne peut avoir deux objets formels differens: or si la conoissance d'un objet extérieur étoit conoissance d'elle même, elle auroit deux objets formels differens, sçavoir l'objet extérieur directement connu, et de plus elle-même pour objet intérieur connu par réflexion; et par conséquent la conoissance d'un objet extérieur, ne peut être conoissance d'elle-même. (Duhamel 1692, chap. 14)

60. Duhamel 1692, chap. 14.

61. "Il est vray que quand je dis je connois, je pense, ce je suppose mon existence . . . ; mais cela n'empesche pas que je ne puisse dire sans contradiction que dans cette proposition *je pense* ce *je* signifie la pensée avant qu'il signifie l'existence, par la raison que je connois l'existence par la pensée, et que je ne connois pas reciproquement la pensée par l'existence; ce qui suffit pour eviter une petition de principe qui consiste à prouver une chose par elle-mesme, considerée en la mesme maniere" (Régis 1692, chap. 13).

62. Here is Régis's full text:

> J'avouë que les idées ne sont pas moins relative essentiellement à leurs objets qu'à leurs sujets; mais avec cette difference, que la relation qu'ils ont avec leurs objets considerez

conclusion of *cogito* does not follow necessarily its premises, given that God can make something think and not exist while it is thinking, by denying that God can do such a thing.[63] On the question of whether the *cogito* is the first proposition known, he simply referred the reader to his previous response to Huet (as he did for all questions about Cartesian doubt).[64]

But we do not have to continue. Rather, we should ask: what does this all signify? Are these all simply atrocious arguments from awful philosophers? Perhaps, but most of these arguments have been repeated, in some guise or another, by philosophers in the twentieth century. More important, we should recognize that these scholastic philosophers are issuing the same basic set of arguments, whether we are dealing with the inquisitors of Angers, Huet, or Duhamel. (One might add Gassendi, in part, and perhaps Bourdin.) All these critics reject the *cogito* because they think of it as a bad argument: it is an enthymeme or it begs the question. None of them can bring themselves to think of the *cogito* as the first principle of knowledge.[65] It just does not look like a first principle; it requires other knowledge (or major premises) that themselves look more like first principles. None of them can think of the *cogito* as the simple intuition of one's existence, a moment in the process of doubt. They all share the same basic criticism.

If one steps back for a moment and looks more closely at these critics, one finds out that they have something else in common. Though not in exactly the same way, all of them reject the method of doubt as a path to certainty. We have already cited the Anger dismissal of Cartesian doubt—it leads to atheism or at least to heresy—and Bourdin is legendary for having

---

entant qu'existans, n'est que contingente & accidentelle. Car il arrive souvent que nous avons des idées dont l'objet n'existe pas actuellement, et comme l'on dit *a parte rei:* au lieu que la relation de nos connoissances à leurs sujets actuellement existans est necessaire et absoluë, n'estant pas possible de concevoir, qu'une connoissance existe separée d'un sujet qui connoit actuellement, et qui est par consequent existant. Ainsi ce n'est par merveille, si voulant déduire mon existence de l'existence de mes connoissances, j'ay plûtot consideré mes connaissances par rapport à leur sujet, que par rapport a leur objet; puisque le rapport qu'elles ont avec celuy-cy, consideré comme existant, n'est que contingent et accidentel, et que le rapport qu'elles ont avce l'autre, est absolu et necessaire. (Régis 1692, chap. 13)

63. "Les cartesiens n'ont point étably de principe duquel il s'ensuive que Dieu puisse faire que je pense et que je ne soit pas: il est vray que j'ai dit . . . qu'il n'y a point d'impossibilité avant le decret de Dieu, mais cela ne veut pas dire que Dieu puisse faire les choses absolument impossibles, comme, *que je pense et que je ne suis pas,* tandis que je pense" (Régis 1692, chap. 13).

64. "Comme les raisons que M. du Hamel; apporte pour prouver que cette proposition: *je pense, donc je suis,* n'est pas la premiere proposition, sont les mesmes que l'Auteur de la Censure de la philosophie Cartésienne a proposées dans le 7 art. du 1. chap. M. du Hamel nous premettra de la renvoyer à la Réponse qui a esté faite sur cet article" (Régis 1692, chap. 18).

65. We should note that Descartes replied to this criticism in advance; see the letter to Clerselier of June or July 1646, AT 4:444–45.

written the lengthiest set of Objections,[66] most of which is directed against the method of doubt: "The method is faulty in its principles ... in the implements it uses ... because it is deficient. ... The method goes astray by failing to reach its goal ... by being excessive ... through negligence ... willfully," etc.[67] Duhamel's rejection of the method of doubt is exemplary:

> The Cartesians pretend to distinguish themselves from the Pyrrhonists in that they do not want to doubt for the sake of doubting, but to be certain, after a sufficient examination, of things about which they have doubts; instead the Pyrrhonists doubt for the sake of doubting, without ever being certain of anything.
>
> But it is clear that, once one doubts everything seriously and effectively, it is impossible to be certain of anything, whatever examination one might conduct, because, if one could be certain of something after such a serious doubt, it would be only by the evidence of the thing, since there is no other rule of human certainty other than the evidence of the thing, according to the Cartesians; now we suppose that they seriously doubt the most evident things, even their own thought and their own existence, and that consequently, it is clear that, after such a general and serious doubt, it would be impossible to be certain of anything, whatever examination one might conduct.
>
> That is why the Cartesians are to be distinguished from the Pyrrhonists in that they do not reason soundly when they say that after a general doubt one can be certain of something, whereas the Pyrrhonists reason soundly and in conformity with their principles when they say that we cannot be certain of anything after having doubted everything.[68]

Huet, in contrast, rejects the method of doubt because he *is* a genuine skeptic, and would rather just wallow in doubt; yet he echoes a similar sentiment: "Descartes and the skeptics believed that we must doubt; but Descartes stopped doubting when it was most necessary to doubt, namely with this principle, *I think, therefore I am*, which is not any less uncertain than all the other things that led him to doubt. The Skeptics continue to doubt this principle and believe that they have many reasons to doubt it. Descartes could not have reproached them if he knew their reason, which is that nothing appears clear enough to them to be admitted as true."[69]

However, without the method of doubt, the *cogito* becomes just another syllogistic argument—and a mediocre one at that.

---

66. Other than Gassendi's separately published *Disquisitio Metaphysica*. For more on Bourdin's *Seventh Objections*, see Ariew 1994b and 1995.
67. AT 7:527–36; CSM 2:358–65. See Ariew 1995.
68. Duhamel 1692, chap. 4; also chaps. 1–3.
69. Huet 1689, chap. 1, art. 14.

In general, what is being criticized by all these scholars is the statement from the *Discourse on Method,* "I think, therefore I am," rather than the pronouncement of Meditation II: "this proposition, *I am, I exist,* is necessarily true whenever it is put forward by me or conceived in my mind."[70]

It does seem that the method of doubt and the *cogito* go hand in hand. The seventeenth-century Scholastics who reject it reject the *cogito* as well. And, of course, a real skeptic who embraces genuine doubt would reject the *method* of doubt (that is, the ability to examine things that one doubts, even the most evident things, and to become certain about them). Such a skeptic, Huet for example, would naturally also reject the *cogito.* Ultimately, the case of Huet should prevent us from speaking too readily of a crucial divide between Scholastics, with their metaphysico-theological leanings, and moderns, with their interest in doubt and self-awareness. But we already knew that such a dichotomy would be simplistic, since many of the figures we count as moderns share what we perceive as the Scholastics' metaphysico-theological leanings. We rank Leibniz and Spinoza as modern. They were endlessly fascinated and repelled by Descartes's philosophy, but they were not much interested in doubt and the *cogito.*[71]

---

70. AT 7:25; CSM 2:17.
71. For Leibniz on the *cogito,* see notes 45 and 56 above. There is the following interesting representation of the *cogito* in Spinoza's *Descartes's Principles of Philosophy,* Prolegomena:

> In whatever direction Descartes turns in order to doubt, he is forced to exclaim, "*I doubt, I think, therefore I am.*" This truth discovered, he finds at the same time the foundations of all the sciences as well as the measure and rule of all other truths, namely: *whatever is as clearly and distinctly perceived as this is true.* That there can be no other foundation of the sciences than this, is more than sufficiently evident from the preceding. For we can call all the rest in doubt with no difficulty, but we cannot doubt this in any way. Concerning this principle, *I doubt, I think, therefore I am,* it should be noted in the first place that it is not a syllogism in which the major premise is omitted. If it were, the premises ought to be clearer and better known than the conclusion, *"therefore I am."* And if this were so, "*I am"* would not be the first foundation of all knowledge. Moreover, it would not be a certain conclusion, for its truth depends upon universal premises which the author had called in question. Therefore *"I think, therefore I am"* is a single proposition equivalent to

For a discussion of the issue of Descartes's modernity and of ancient and modern philosophers, see Sorell 1994, 29–47; Des Chene 1995b.

Appendix: Gilson's *Index* Indexed

This is an index of the entries in Etienne Gilson's *Index-scolastico cartésien* (Gilson 1913). The numbers under the headings are simply the paragraph numbers given by Gilson in the *Index*.

Term	Thomas	Coimbrans	Toletus	Rubius	Suarez	Eustachius
abstraction						1
accident					3, 5	2, 4, 6
acte			7			
action	8, 12		11		9, 10	
affectus	13					
air	16		14, 15			
alchimie		17				
altération		18				
âme	21, 23, 24, 26, 30–31, 33–35	25, 29, 27–29, 32				19, 22
amour			36			
ange	37–39					43
appétit	40–42					
arc-en-ciel		44, 45				
assentiment	46–49					
astrologie	50					
atome					51	
autorité				52		
béatitude						53
bien	57					54–56
categorie						58
cause	66, 68, 74	59		67	60–61, 65, 69, 72–73	62–64, 70–71
chose			75			
cieux		76				
comète		77				
composé						78
concept			79		80	
concours			81			
condensation			82–83			
connaissance	90–91	86–87, 92	88–89			84–85

[207]

APPENDIX

Term	Thomas	Coimbrans	Toletus	Rubius	Suarez	Eustachius
continu			93–96			
corps	102		99–101			97–98
corruption		103				
couleur		104–5				
couronne		106				
création	107–11					112
definition						113
délibération	114					
démonstration			117			115–16
dénomination				118		
dialetique						119
Dieu	120, 123–26, 128–36, 138–41			137	121, 127	122
dispute	146					
distinct		147				148
distinction					148b, 150, 152–53	149, 151
election	154–55					
éléments		156–58				
énconciation			159			
entendement			160–61			
entité						162
équivoque	163					
erreur			164–67			
espèce		168				169–70
esprits animaux		171–76				
essence	179				180	177–78
étendue, extension			181			
être	185–86			184, 187, 190–91	183	182, 188–89, 192
faim		193–94				
falsitas, fausseté	195				195b	
feu		196–97				
figure	198					
fin	199					
fixe		200–203				
flux						204
foi	205–7					
fontaine						208
forme	209	212	210		211	
foudre		213				
genre						214–15
habitus	218					216–17
humeur						219
idée	220–23					
identité						224
imagination		228				225–27
immortalité		228b–229				
infini	231–32	233, 236–37			234–35	230, 238
inné	239–43					
instinct						244
jugement	245					
liberté, libre arbitre	247–48, 251				249	250
lieu	252		253			254
lumière	257, 259, 261	255, 258			260	256

Gilson's Index Indexed

Term	Thomas	Coimbrans	Toletus	Rubius	Suarez	Eustachius
magie		262				
mal	263, 265–67				264	
manne		268				
mathématique	270	269, 271			274	
matière		272–73, 275				
métempsychose		281				
météores		283–84				
méthode						285
monde	287–88	288b	289			286
mouvement	297	290–92, 296, 300–301, 308	293–94, 299, 302–4, 307			295, 298, 305–6
nature		309, 313–14	310, 312		311	
notion		316				317–18
nue		319				
parélie		320				
passion	323–33, 335–42, 344–45	322, 334, 343				321
péché			346			
pensée	348					347
perfection	349–53	354				
pesanteur		355				
pétition de principe	356					
philosophie		357, 360–62			359	358
physique		363				
plaisir	354					
pluie		365				
possible	366					
pouvoir				367		
précipitation	368–70					
prédicat						371
préjugé	372					
présage		373–74				
principe		378–80			375	376–77
privation	383				382	381
propre, propriété						384
providence	385					
puissance	388	387, 389–90			386	
qualité	391		393			392
quantité	394–95	396, 401	399	398, 400		397
raison	402		403			
réminiscence	404	405				
rire		406				
sagesse						407
science						408
sel						
sens, sens commun	409	414–16, 419	418			410–13, 417
sensible						420
sommeil		421–23				
son		424–25				
substance	426–27		435	433, 436	431–32, 434	428–30
suppositum						437
surface			438	439		
syllogisme						440–41

[210] APPENDIX

Term	Thomas	Coimbrans	Toletus	Rubius	Suarez	Eustachius
tempérament						442
temps	445, 447	443–44			446	
terre		448–51				
transubstantation	452–57					
Trinité	458					
union	459–61, 463	462, 464–67				468
unité						469
universeaux			473			470–72
univoque	475–76	477–78				474
vapeur		479				
vent		480				
vérité						481–84
vertu	485–90					
vide		491–92				495
violent		492b–94				
volonté	500–501					496–99
vue		502–6				507
Total						
$N = 511$	168 (33 %)	135 (26 %)	50 (10 %)	14 (3 %)	37 (7 %)	107 (21 %)

Note: *In his Appendix to the* Index, *Gilson also has references to Chauvin, numbered 508–25.*

Bibliography

A., M. G. de L'[Pierre Daniel Huet]. 1692. *Nouveaux mémoires pour servir à l'histoire du cartésianisme.* n.p, n.d.
Actes du Congrès du Tricentenaire de Pierre Gassendi. 1957. Digne.
Albert of Saxony. 1567. *Quaestiones super quatuor libros de caelo et mundo.* Rome.
Alexander, H. G., ed. 1956. *The Leibniz-Clarke Correspondence.* Manchester: Manchester University Press.
Amici, Bartolomeo d'. 1626. *In Aristotelis libros De caelo et mu[n]do : dilucida textus explicatio et disputationes, in quibus illustrium scholarum Averrois, D. Thomae, Scoti et nominalium sententiae.* . . . Naples.
Aquinas, Thomas. 1918–30. *Summa contra gentiles.* Rome.
———. 1933. *Opusculum De ente et essentia.* Ed. C. Boyer. Rome: Gregorian University Press.
———. 1945. *Introduction to St. Thomas Aquinas.* Trans. Anton C. Pegis. New York: Random House.
———. 1953. *In octo libros De physico auditu sive physicorum Aristoteis commentaria.* Ed. P. Angeli and M. Pirotta. Naples: M. D'auria.
———. 1954. *De Natura Loci.* Opusc. LII of *Opuscula philosophica.* Ed. M. Spiazzi. Taurini: Marietti.
———. 1964–76. *Summa theologiae.* Oxford: Blackfriars.
Argentré, Duplessis d'. 1736. *Collectio judiciorum de novis erroribus tomus tertium.* Paris.
Ariew, Roger. 1984. "Galileo's Lunar Observations in the Context of Medieval Lunar Theory." *Studies in History and Philosophy of Science* 15, no. 3: 213–26.
———. 1986. "Descartes as Critic of Galileo's Scientific Methodology." *Synthese* 67: 77–90.
———. 1987a. "The Infinite in Descartes' Conversation with Burman." *Archiv für Geschichte der Philosophie* 69: 140–63.
———. 1987b. "The Phases of Venus before 1610." *Studies in History and Philosophy of Science* 18: 81–92.
———. 1992. "Descartes and the Tree of Knowledge." *Synthese* 92: 101–16.
———. 1994a. "Quelques condamnations du cartésianisme: 1662–1706." *Bulletin cartésien XXII, Archives de Philosophie* 57: 1–6.

———. 1994b. "Sur les septièmes réponses." In *Descartes: Objecter et répondre*, ed. Jean-Marie Beyssade and Jean-Luc Marion, 123–40. Paris: P.U.F.
———. 1995. "Pierre Bourdin and the Seventh Objections." In *Descartes and His Contemporaries: Meditations, Objections, and Replies*, ed. Roger Ariew and Marjorie Grene, 208–25. Chicago: University of Chicago Press.
———. 1996. "Les *Principia* et la *Summa philosophica quadripartita*." In *Descartes Principia Philosophiae (1644–1994)*, ed. Jean-Robert Armogathe and Giulia Belgioioso, 473–90. Naples: Vivarium.
———. 1998. "Aristotelianism in the 17th Century." In *Routledge Encyclopedia of Philosophy*, ed. E. Craig, 1:386–93. London: Routledge.
———. Forthcoming. "Descartes and the Jesuits." In *The Jesuits and the Scientific Revolution*, ed. Mordechai Feingold. Princeton: Princeton University Press.
Ariew, Roger, John Cottingham, and Tom Sorell, eds. and trans. 1998. *Descartes' Meditations: Background Source Materials*. Cambridge: Cambridge University Press.
Ariew, Roger, and Marjorie Grene, eds. 1995. *Descartes and His Contemporaries*. Chicago: University of Chicago Press.
Aristotle. 1910–52. *The Works of Aristotle*. Ed. W. D. Ross. Oxford.
Armogathe, Jean-Robert. 1977. *Theologia cartesiana: l'explication physique de l'Eucharistie chez Descartes et dom Desgabets*. The Hague: M. Nijhoff.
———. 1994. "L'approbation des *Meditationes* par la faculté de théologie de Paris (1641)." *Bulletin Cartésien XXI*, *Archives de Philosophie* 57: 1–3.
Armogathe, Jean-Robert, and Giulia Belgioioso, eds. 1996. *Descartes: Principia Philosophiae (1644–1994)*. Naples: Vivarium.
Armour, Leslie. 1988. "Le cartésianisme au Québec." *Bulletin Cartésien XVI*, *Archives de Philosophie* 51: 1–12.
———. 1993. "Descartes and Eustachius a Sancto Paulo: Unravelling the Mind-Body Problem." *British Journal for the History of Philosophy* 1: 3–21.
Arnauld, Antoine. 1775. *Oeuvres*. Paris.
———. 1964. *The Art of Thinking*. Trans. James Dickoff and Patricia James. Indianapolis: Bobbs-Merrill.
Arriaga, Rodericus. 1632. *Cursus philosophicus*. Antwerp.
Aubrey, John. 1898. *Brief Lives, Chiefly of Contemporaries, between the Years 1669 and 1696*. Oxford.
Babin, François. 1679. *Journal ou relation fidele de tout ce qui s'est passé dans l'université d'Angers au sujet de la philosophie de Des Carthes en l'execution des ordres du Roy pendant les années 1675, 1676, 1677, et 1678*. Angers.
Baillet, Adrien. 1691. *La vie de monsieur Descartes*. Paris.
Barbay, Pierre. 1675–76. *Commentarius in Aristotelis logicam, metaphysicam, moralem, physicam*. Paris.
Barker, Peter, and Bernard R. Goldstein. 1988. "The Role of Comets in the Copernican Revolution." *Studies in History and Philosophy of Science* 19: 299–319.
Basso, Sebastian. 1649 [1621]. *Philosophiae naturalis adversus Aristotelem. In quibus abstrusa veterum physiologia restauratur, et Aristotelis errores solidis rationibus refelluntur*. 2d ed. Amsterdam.
Baumgartner, Frederic J. 1987. "Sunspots or Sun's Planets: Jean Tarde and the Sunspot Controversy of the Early Seventeenth Century." *Journal for the History of Astronomy* 18: 44–54.
Bayle, Pierre. 1684. *Recueil de quelques pièces curieuses concernant la philosophie de Monsieur Descartes*. Amsterdam.

Beck, L. J. 1965. *The Metaphysics of Descartes*. Oxford: Clarendon Press.
Bellarmine, R. 1984. *Louvain Lectures*. Trans. U. Baldini and G. V. Coyne. *Studi Galileiani* 1.
Bernier, François. 1682. *Doutes sur quelques-uns des principaux chapitres de son abrégé de la philosophie de Gassendi*. Paris.
———. 1684. *Abrégé de la philosophie de Gassendi*. 6 vols. 2d ed. Lyon.
Bettoni, Efrem. 1961. *Duns Scotus: The Basic Principles of His Philosophy*. Washington, D.C.: Catholic University of America Press.
Blair, Ann. 1993. "The Teaching of Natural Philosophy in Early Seventeenth-Century Paris: The Case of Jean-Cecile Frey." *History of Universities* 12: 95–158.
———. 1994. "Tradition and Innovation in Early Modern Natural Philosophy: Jean Bodin and Jean-Cecile Frey." *Perspectives on Science* 2: 428–54.
Bloch, Olivier R. 1971. *La philosophie de Gassendi*. The Hague: Martinus Nijhoff.
Blum, Paul Richard. Forthcoming. "Aristotelianism 'More Geometrico': The Case of Honoré Fabri." In *Conversations with Aristotle*, ed. C. Blackwell and S. Kusukawa. Dordrecht: Kluwer.
Boccafuoco, Costanzo. 1589. *Conciliatio dilucida omnium controversiarum, quae in doctrina duorum summorum theologorum S. Thomae, & subtilis Ioannis Scoti passim leguntur. . . .* Rome.
Boileau. 1747. *Oeuvres de Boileau*, ed. Saint-Marc. Paris.
Bonaviolensis, Petrus Iaubert. 1657. Student thesis, Collegio Cadurcensi Societatis Jesu. Bibliothèque Nationale, Cabinet d'Estampes AA6.
Boot, Arnold de, and Gerard de Boot. 1641. *Philosophia naturalis reformata*. Dublin.
Borbonius, Antonius. circa 1622. Student thesis, Collegio Paris. Claremontano Societ. Jesu. Bibliothèque Nationale, Cabinet d'Estampes AA6.
Bossu, René le. 1981 [1674]. *Parallele des principes de la physique d'Aristote et de celle de René Des Cartes*. Reprint. Paris: Vrin.
Bouillier, Francisque. 1868. *Histoire de la philosophie cartésienne*. 2 vols. Paris.
Bouju, Théophraste. 1614. *Corps de toute la philosophie*. Paris.
Bourdin, Pierre. 1639. *Prima geometriae elementa*. Paris.
———. 1640. *Geometria, nova methodo*. Paris.
———. 1643. *L'introduction à la mathématique contenant les coignaissances, et pratiques necéssaires à ceux qui commencent d'apprendre les mathématiques. . . .* Paris.
———. 1646. *Sol flamma sive tractatus de sole, ut flamma est, eiusque pabulo sol exurens montes, et radios igneos exsufflans Eccles. 43. Aphorismi analogici parvi mundi ad magnum magni ad parvuum*. Paris.
———. 1655a. *L'architecture militaire ou l'art de fortifier les places regulières et irregulières*. Paris.
———. 1655b. *Le dessein ou la perspective militaire*. Paris.
———. 1661. *Cours de mathématique*. 3d ed. Paris.
Brockliss, L. W. B. 1981. "Aristotle, Descartes, and the New Science: Natural Philosophy at the University of Paris, 1600–1740." *Annals of Science* 38: 33–69.
———. 1987. *French Higher Education in the Seventeenth and Eighteenth Centuries: A Cultural History*. Oxford: Clarendon Press.
———. 1992. "The Scientific Revolution in France." In *The Scientific Revolution in National Context*, ed. R. S. Porter and M. Teich, 55–89. Cambridge: Cambridge University Press.
———. 1995a. "Descartes, Gassendi and the Reception of the Mechanical Philosophy in the French Collèges de Plein Exercice, 1640–1730." *Perspectives on Science* 3: 450–79.

———. 1995b. "Pierre Gautruche et l'enseignement de la philosophie de la nature dans les collèges jésuites français vers 1650." In *Les Jésuites à la Renaissance. Système éducatif et production du savoir*, ed. L. Giard, 187–219. Paris: Presses Universitaires de France.
Buchdahl, Gerd. 1969. *Metaphysics and the Philosophy of Science*. Oxford: Basil Blackwell.
Burgersdijk, Franco. 1657. *Institutionum metaphysicarum*. The Hague.
Buridan, Jean. 1509. *Questiones super octo phisicorum libros Aristotelis*. Paris.
Capek, Milic. 1976. *The Concepts of Space and Time*. Dordrecht: Reidel.
Carraud, Vincent. Forthcoming. "La matière assume toutes les formes: note sur le concept d'ordre et sur une proposition thomiste de la cosmogonie cartésienne." In *La Conception de la Science*, ed. J. Gayon.
Carriero, John Peter. 1990. *Descartes and the Autonomy of Human Understanding*. New York: Garland.
Casimir de Toulouse. 1674. *Atomi peripateticae, sive tum veterum tum recentiorum atomistarum placita ad neotericae scholae methodum redacta*. Toulouse.
Ceriziers, René de. 1643. *Le philosophe français*. Paris.
Chevreul, Jacques du. 1623. *Sphaera*. Paris.
———. 1624. *Commentarius in libros meteorologicos*. Ms. Cherbourg 23.
Clave, Etienne de. 1641. *Nouvelle lumiere philosophique*. Paris.
Clavius, Christopher. 1611–12. *Opera mathematica*. Rome.
Conimbricenses. 1598. *Commentarii in tres libros de anima*. Coimbra.
———. 1606. *Commentarii in universam dialecticam Aristotelis*. Coimbra.
———. 1986 [1592]. *Commentarii in octo libros physicorum Aristotelis*. Reprint. Hildesheim: Olms.
Cottingham, John, ed. 1991. *The Cambridge Companion to Descartes*. Cambridge: Cambridge University Press.
Cousin, Victor. 1866. *Fragments philosophiques pour servir à l'histoire de la philosophie*. Paris.
Crassot, Jean. 1618. *Physica*. Paris.
Cusa, Nicholas. 1962 [1514]. *Opera*. Frankfurt: Minerva.
Dainville, Francois de. 1987. *L'éducation des Jésuites*. Paris: Editions de Minuit.
Dalbiez, Roland. 1929. "Les sources scolastiques de la théorie cartésienne de l'être objectif. A propos du 'Descartes' de M. Gilson." *Revue d'Histoire de la Philosophie* 3: 464–72.
Daniel, Gabriel. 1690. *Voyage du monde de Descartes*.
———. 1692. *A Voyage to the World of Cartesius*. London.
———. 1693. *Nouvelles difficultés proposées par un péripatéticien à l'auteur du "Voyage du monde de Descartes."* Paris.
Dear, Peter. 1988. *Mersenne and the Learning of the Schools*. Ithaca, N.Y.: Cornell University Press.
Descartes, René. 1811. *Pensées de Descartes sur la religion et la morale*. Paris.
———. 1964–74. *Oeuvres de Descartes*. Ed. C. Adam and P. Tannery. 2d ed. Paris: Vrin.
———. 1979. *The World/Le Monde*. Trans. M. S. Mahoney. New York: Abaris.
———. 1983. *Principles of Philosophy*. Trans. V. R. Miller and R. P. Miller. Dordrecht: Reidel.
———. 1984–91. *The Philosophical Writings of Descartes*. Vols. 1 and 2, trans. J. Cottingham, R. Stoothoff, and D. Murdoch. Vol. 3, trans. J. Cottingham, R. Stoothoff, D. Murdoch, and Anthony Kenny. Cambridge: Cambridge University Press.
Des Chene, Dennis. 1995a. "Cartesiomania: Early Receptions of Descartes." *Perspectives on Science* 3: 534–81.

———. 1995b. *Physiologia: Philosophy of Nature in Descartes and the Aristotelians*. Ithaca, N.Y.: Cornell University Press.
Dictionnaire de l'Académie française. 1765. Paris.
Di Vona, Piero. 1994. *I concetti trascendenti in Sebastian Izquierdo e nella Scolastica del seicento*. Napoli: Loffredo.
Douarche, Aristide. 1970 [1888]. *L'Université de Paris et les Jésuites*. Geneva: Slatkine Reprints.
Drake, Stillman. 1957. *Discoveries and Opinions of Galileo*. New York: Doubleday.
———. 1978. *Galileo at Work*. Chicago: University of Chicago Press.
Drake, Stillman, and C. D. O'Malley, trans. 1960. *The Controversy on the Comets of 1618*. Philadelphia: University of Pennsylvania Press.
Duhamel, Jean. 1692. *Reflexions critiques sur le système cartesien de la philosophie de mr. Régis*. Paris.
———. 1705. *Philosophia universalis, sive commentarius in universam Aristotelis philosophiam, ad usum scholarum comparatam*. 5 vols. Paris.
Duhamel, Jean-Baptiste. 1660. *Astronomia Physica*. Paris.
———. 1677. *Philosophia vetus et nova*. Amsterdam.
Duhem, Pierre. 1908–13. *Etudes sur Léonard de Vinci*. 3 vols. Paris: Hermann.
———. 1913–59. *Le Système du Monde*. 10 vols. Paris: Hermann.
———. 1985. *Medieval Cosmology*. Ed. and trans. Roger Ariew. Chicago: University of Chicago Press.
Dupleix, Scipion. 1627. *Corps de philosophie, contenant la logique, l'ethique, La physique, et la metaphysique*. Geneva.
———. 1984. *La logique*. Paris: Fayard.
———. 1990. *La physique*. Ed. Roger Ariew. Paris: Fayard.
———. 1992. *La métaphysique*. Ed. Roger Ariew. Paris: Fayard.
———. 1993. *L'ethyque*. Ed. Roger Ariew. Paris: Fayard.
Echarri, J. 1950. "Uno influjo español desconcido en la formación del sistema cartesiano; dos textos paralelos de Toledo y Descartes sobre el espacio." *El Pensamiento* 6: 291–332.
Emerton, Norma E. 1984. *The Scientific Reinterpretation of Form*. Ithaca, N.Y.: Cornell University Press.
Eustachius a Sancto Paulo. 1613–16. *Summa theologiae tripartita*. Paris.
———. 1629 [1609]. *Summa philosophiae quadripartita*. Cologne.
Fabri, Honoré. 1666. *De plantis et de generatione animalium, de homine*. Lyons.
Fattori, M., ed. 1997. *Il vocabulario della republique des lettres*. Florence: Leo Olschki.
Fattori, M., and M. L. Bianchi, eds. 1990. *Idea*. Rome: Edizioni dell'Ateno.
Faye, Emmanuel. 1986. "Le corps de philosophie de Scipion Dupleix et l'arbre cartésien des sciences." *Corpus* 2: 7–15.
Frassen, Claude. 1668. *Philosophia academica, quam ex selectissimis Aristotelis et Doctoris Subtilis Scoti rationibus*.... Paris.
———. 1677. *Scotus academicus*. Paris.
Freedman, Joseph. 1993. "The Diffusion of the Writings of Petrus Ramus in Central Europe, c. 1570–c. 1630." *Renaissance Quarterly* 46: 98–152.
Frey, Jean-Cecile. 1628. *Cribrum philosophorum qui Aristotelem superiore et hac aetate oppugnarunt*. Paris.
———. 1633. *Universae philosophiae compendium*. Paris.
Furetière, Antoine. 1979 [1690]. *Dictionnaire universel*. Reprint. Geneva: Slatkins.

Galileo Galilei. 1890–1901. *Opere.* Ed. A. Favaro. Florence.
——. 1974. *Two New Sciences.* Trans. S. Drake. Madison: University of Wisconsin Press.
Garber, Daniel. 1988. "Descartes, the Aristotelians and the Revolution That Did Not Happen in 1637." *Monist* 71: 471–86.
——. 1992. *Descartes's Metaphysical Physics.* Chicago: University of Chicago Press.
——. 1994. "On the Front Lines of the Scientific Counter-Revolution: Defending Aristotle Paris-Style." Lecture delivered at the conference "Tradition and Novelty: Cultural and Regional Considerations in the Competition between the Old and the New Science in the Seventeenth Century." Chicago Humanities Institute, 1994.
Garber, Daniel, and Lesley Cohen. 1982. "A Point of Order: Analysis, Synthesis, and Descartes's *Principles.*" *Archiv für Geschichte der Philosophie* 64: 136–47.
Garber, Daniel, and Michael Ayers, eds. 1997. *Cambridge History of Seventeenth Century Philosophy.* 2 vols. Cambridge: Cambridge University Press.
Garnier, Jean. 1651. Physics course, Collège de Clermont, Paris. Bibliothèque Nationale, Fond Latin, ms. 11257.
Gassendi, Pierre. 1624. *Exercitationes paradoxicae adversus Aristoteleos.* Paris.
——. 1642. *De motu impresso a motore translato.* Paris.
——. 1959 [1624]. *Dissertation en forme de paradoxes contre les Aristotéliciens.* Ed. and trans. B. Rochot. Paris: Vrin.
——. 1962. *Disquisitio Metaphysica.* Ed. and trans. B. Rochot. Paris: Vrin.
Gaukroger, Stephen. 1994. "The Sources for Descartes's Procedure of Deductive Demonstration in Metaphysics and Natural Philosophy." In *Reason, Will, and Sensation: Studies in Descartes's Metaphysics,* ed. John Cottingham, 47–62. Oxford: Clarendon Press.
Gaultruche, Pierre. 1656. *Institutio totius mathematicae.* Caens.
——. 1665. *Philosophiae ac mathematicae totius clara, brevis, et accurata institutio.* 5 vols. Caen.
Gilson, Etienne. 1913. *Index scolastico-cartésien.* Paris: Félix Alcan.
——. 1925. *Discours de la méthode, texte et commentaire.* Paris: Vrin.
——. 1930. *Etudes sur le rôle de la pensée médievale dans la formation du système cartésien.* Paris: Vrin.
——. 1952. *Jean Duns Scot. Introduction à ses positions fondamentales.* Paris: Vrin.
——. 1989 [1913]. *La liberté chez Descartes et la théologie.* Paris: Vrin.
Goclenius, Rudolphus. 1964 [1613]. *Lexicon philosophorum.* Reprint. Hildesheim: Olms
Goudin, Antoine. 1864 [1st Latin ed. 1668]. *Philosophie suivant les principes de Saint Thomas.* Trans. T. Bourard. Paris.
Gouhier, Henri. 1972. *La pensée religieuse de Descartes.* 2d ed. Paris: Vrin.
Grandamy, Jacques. 1665. *Le cours de la comete qui a paru sur la fin de l'année 1664 et au commencement de l'année 1665. Avec un traité de sa nature, de son mouvement, et ses effets.* Paris.
Grange, Jean-Baptiste de la. 1682 [privilège et permission, 1675]. *Les principes de la philosophie contre les nouveaux philosophes, Descartes, Rohault, Régius, Gassendi, le p. Maignan, etc. Vol. 1, Traité des qualitez; Vol. 2, Traité des éléments et météores.* Paris.
Grant, Edward. 1976. "Place and Space in Medieval Physical Thought." In *Motion and Time, Space and Matter,* ed. P. Machamer and R. Turnbull, 137–67. Columbus: Ohio State University Press.
——. 1981. *Much Ado about Nothing.* Cambridge: Cambridge University Press.
——. 1987. "Celestial Orbs in the Latin Middle Ages." *Isis* 78: 153–73.

———. 1996a. *Planets, Stars, and Orbs*. Cambridge: Cambridge University Press.
———. 1996b. *The Foundations of Modern Science in the Middle Ages*. Cambridge: Cambridge University Press.
Grene, Marjorie. 1985. *Descartes*. Minneapolis: University of Minnesota Press.
———. 1991. *Descartes among the Scholastics*. Milwaukee: Marquette University Press.
———. 1993. "Aristotelico-Cartesian Themes in Natural Philosophy: Some Seventeenth-Century Cases." *Perspectives on Science* 1: 66–87.
———. 1995. "Animal Mechanism and the Cartesian Vision of Nature." In *Physics, Philosophy, and the Scientific Community*, ed. K. Gavroglu et al., 189–204. Dordrecht: Kluwer.
Gueroult, Martial. 1968. *Descartes selon l'ordre des raisons*. 2 vols. Paris: Aubier Montaigne.
———. 1984–85. *Descartes' Philosophy Interpreted according to the Order of Reasons*. Trans. Roger Ariew, with the assistance of Robert Ariew and Alan Donagan. 2 vols. Minneapolis: University of Minnesota Press.
Heereboord, Adriaan. 1668. *Philosophia Naturalis*. Ed. posthuma. Oxford.
———. 1680. *Meletemata philosophica*. Ed. nova. Amsterdam.
Heilbron, John L. 1979. *Electricity in the 17th and 18th Centuries*. Berkeley: University of California Press.
Hellyer, Marcus. 1996. "'Because the Authority of My Superiors Commands': Censorship, Physics, and the German Jesuits." *Early Science and Medicine* 3: 319–54.
Heyd, Michael. 1982. *Between Orthodoxy and the Enlightenment: Robert Chouet and the Introduction of Cartesian Science in the Academy of Geneva*. The Hague: Martinus Nijhoff.
Hintikka, J. 1978. "A Discourse on Descartes's Method." In *Descartes: Critical and Interpretive Essays*, ed. M. Hooker, 74–88. Baltimore: Johns Hopkins University Press.
Hintikka, J., and U. Remes. 1974. *The Method of Analysis*. Dordrecht: Reidel.
Hobbes, Thomas. 1839–45. *The English Works of Thomas Hobbes of Malmesbury*. Ed. William Molesworth. London.
Huet, Pierre Daniel. 1689. *Censura philosophiae cartesianae*. Paris.
Huguet, Edmond. 1949. *Dictionnaire de la langue française du seizième siècle*. Paris: Didier.
John of St. Thomas. 1663. *Cursus philosophicus Thomisticus, secundum exactam, veram et genuinam Aristotelis. . . .* Lyon.
Jolley, Nicholas, ed. 1995. *The Cambridge Companion to Leibniz*. Cambridge: Cambridge Universiy Press.
Jones, P. J. 1947. "The Identity of the Author of a Hitherto Anonymous Work." *Scripta Mathematica* 13: 119–20.
Jordan, Mark D. 1984. "The Intelligibility of the World and the Divine Ideas in Aquinas." *Review of Metaphysics* 38: 17–32.
Koyré, Alexandre. 1957. *From the Closed World to the Infinite Universe*. Baltimore: Johns Hopkins University Press.
———. 1978. *Galileo Studies*. Trans. J. Mepham. New Jersey: Humanities Press.
Kubbinga, H. H. 1984. "Les premières théories 'moléculaires': Isaac Beeckman (1620) et Sébastien Basson (1621)." *Revue d'Histoire des Sciences et de leurs applications* 37: 215–33.
Lattis, James M. 1989. *Christopher Clavius and the SPHERE of Sacrobosco: The Roots of Jesuit Astronomy on the Eve of the Copernican Revolution*. Ph.D. diss., University of Wisconsin.

———. 1994. *Between Copernicus and Galileo: Christoph Clavius and the Collapse of Ptolemaic Cosmology.* Chicago: University of Chicago Press.
Laudan, Larry. 1981. *Science and Hypothesis.* Dordrecht: Reidel.
Launoy, Jean de. 1656. *De varia Aristotelis fortuna.* 2d ed. Hages-Comitum.
Le Grand, Antoine. 1972 [1694]. *Entire Body of Philosophy according to the Principles of the Famous Renate Descartes.* Reprint. New York: Johnson Reprints.
Leibniz, G. W. 1989. *Philosophical Essays.* Ed. and trans. R. Ariew and D. Garber. Indianapolis: Hackett.
———. 1992. *De Summa Rerum.* Trans. G. H. R. Parkinson. New Haven: Yale University Press.
Leijenhorst, Cees. 1996. "Jesuit Concepts of *Spatium Imaginarium* and Thomas Hobbes's Doctrine of Space." *Early Science and Medicine* 1. 355-80.
Lennon, Thomas M. 1993. *The Battle of the Gods and Giants.* Princeton: Princeton University Press.
Lennon, Thomas M., and Patricia Ann Easton. 1992. *The Cartesian Empiricism of François Bayle.* New York: Garland.
Lennon, Thomas, J. W. Nicholas, and J. W. Davis, eds. 1982. *Problems of Cartesianism.* Kingston: McGill-Queen's University Press.
Lenoble, R. 1943. *Mersenne ou la naissance du mécanisme.* Paris: Vrin.
Lesclache, Louis de. 1651. *La philosophie en tables, divisée en cinq parties.* Paris.
Lewis, Eric P. 1995. *Descartes and Tradition: The Miracle of the Eucharist.* M.A. Thesis, Virginia Polytechnic Institute and State University.
Llamazares, Tomas de. [1669?]. *Cursus philosophicus: sive, Philosophia scholastica ad mentem Scoti, nova & congruentiore addiscentibus methodo disposita.* Lyon.
Lohr, Charles H. 1974. "Renaissance Latin Aristotle Commentaries: Authors A–B." *Studies in the Renaissance* 21: 228–89.
———. 1975. "Renaissance Latin Aristotle Commentaries: Authors C." *Renaissance Quarterly* 28: 689–741.
———. 1976. "Renaissance Latin Aristotle Commentaries: Authors D–F." *Renaissance Quarterly* 29: 714–45.
———. 1980. "Renaissance Latin Aristotle Commentaries: Authors Pi–Sm." *Renaissance Quarterly* 33: 623–734.
———. 1982. "Renaissance Latin Aristotle Commentaries: Authors So–Z." *Renaissance Quarterly* 35: 164–256.
———. 1988. *Latin Aristotle Commentaries, II: Renaissance Authors.* Florence.
———. Forthcoming. "Latin Aristotelianism and the Seventeenth-Century Calvinist Theory of Scientific Method."
Lüthy, Christoph. 1997. "Thoughts and Circumstances of Sébastien Basson: Analysis, Micro-History, Questions." *Early Science and Medicine* 2: 1–73.
Lux, David S. 1989. *Patronage and Royal Science in Seventeenth-Century France.* Ithaca, N.Y.: Cornell University Press.
Maignan, Emanuel. 1653. *Cursus philosophicus.* Toulouse.
Malbreil, Germain. 1991. "Descartes censuré par Huet." *Revue Philosophique* 3: 311–28.
Mandonnet, Pierre. 1908. *Siger de Brabant et l'averroisme latin au XIIIme siècle.* Louvain.
Marandé, Léonard. 1642. *Abrégé curieux de toute la philosophie.* Paris.
Marion, Jean-Luc. 1975. *Sur l'ontologie grise de Descartes.* Paris: Vrin.
———. 1981. *Sur la théologie blanche de Descartes.* Paris: P. U. F.
———. 1986. *Sur le prisme métaphysique de Descartes. Constitution et limites de l'onto-théologie.* Paris: P.U.F.

Matthews, M. R., ed. 1989. *The Scientific Background to Modern Philosophy.* Indianapolis: Hackett.
McClaughlin, Trevor. 1979. "Censorship and Defenders of the Cartesian Faith in Mid-Seventeenth Century France." *Journal of the History of Ideas* 40: 563–81.
Meinel, Christoph. 1988. "Early Seventeenth-Century Atomism, Theory, Epistemology, and the Insufficiency of Experiment." *Isis* 79: 68–103.
Melsen, Andrew G. van. 1960. *From Atomos to Atom.* New York: Harper.
Mersenne, Marin. 1624. *L'impiété des Déistes, Athées, et Libertins de ce temps, combattue, et renversée de point en point par raisons tirée de la Philosophie, et de la Théologie.* Paris.
——. 1625. *La verité des sciences.* Paris.
——. 1644. *Cogitata physico-mathematica.* Paris, 1644.
——. 1933–88. *La correspondance du P. Marin Mersenne, religieux minime.* Ed. C. de Waad et al. 17 vols. Paris: P.U.F. and C.N.R.S.
Michael, Emily. 1997. "Daniel Sennert on Matter and Form: At the Juncture of the Old and the New," *Early Science and Medicine* 2: 272–99.
Michael, Emily, and Fred S. Michael. 1989. "Corporeal Ideas in Seventeenth-Century Psychology." *Journal of the History of Ideas* 50: 31–48.
Monumenta paedagogica Societatis Jesu. 1901. Matriti: A. Avrial.
Morus, Michael. 1716. *Vera sciendi methodus.* Paris.
——. 1726. *De principiis physicis seu corporum naturalium disputatio.* Paris.
Moulin, Pierre du. 1644. *La philosophie, mise en francois, et divisée en trois parties, scavoir, elements de la logique, la physique ou science naturelle, l'ethyque ou science morale.* Paris.
Murr, Sylvia, ed. 1992. *Corpus* 20/21.
Nadler, Steven. 1988. "Arnauld, Descartes, and Transubstantiation: Reconciling Cartesian Metaphysics and Real Presence." *Journal of the History of Ideas* 59: 229–46.
——. 1989. *Arnauld and the Cartesian Philosophy of Ideas.* Princeton: Princeton University Press.
Nicot, Jean. 1979 [1606]. *Trésor de la langue française tant ancienne que moderne.* Reprint. Paris, Le Temps.
Nielsen, Lauge Olaf. 1988. "A 17th-Century Physician on God and Atoms: Sebastian Basso." In *Meaning and Inference in Medieval Philosophy: Studies in Memory of Jan Pinborg,* ed. Norman Kretzmann, 297–369. Dordrecht: Kluwer.
Normore, Calvin. 1986. "Meaning and Objective Being." In *Essays on Descartes' Meditations,* ed. A. O. Rorty, 223–40. Berkeley: University of California Press.
Oldenburg, Henry. 1966. *The Correspondence of Henry Oldenburg.* Ed. A. R. Hall and M. B. Hall. Madison: University of Wisconsin Press.
Oresme, Nicole. 1968. *Livre du ciel et du monde.* Ed. and trans. A. Menu and D. Denomy. Madison: University of Wisconsin Press.
Oxford English Dictionary. 1989. Oxford: Clarendon Press.
Pachtler, G. M., ed. 1890. *Ratio studiorum et institutiones scholasticae Societatis Jesu per Germaniam olim vigentes,* vol 3. Monumenta Germaniae Paedagogica 9. Berlin: Hofmann.
Panofsky, Erwin. 1968. *Idea, a Concept in Art Theory.* Trans. Joseph J. S. Peake. Columbia: University of South Carolina Press.
Pemble, William. 1650 [1628]. *De formarum origine.* Ed. posthuma. Cambridge.
Poncius, Joannes. 1672. *Philosophiae ad mentem Scoti cursus integer.* Lyon.
Pourchot, Edmond. 1695. *Institutio philosophica.* Paris.
Prou, Ludovicus. 1665. *De hypothesi Cartesiana positiones physico mathematica.* Paris.

Raconis, Charles François d'Abra de. 1651. *Tertia pars philosophiae, seu Physica*. Paris.
Rada, Juan de. 1620. *Controversiae theologiae inter S. Thomam & Scotum, super quatuor libros sententiarum. In quibus pugnantes sententiae referuntur, potiores difficultates elucidantur, & responsiones ad argumenta Scoti reiiciuntur*. . . . Cologne.
Rapin, René. 1725. *Reflexions sur la philosophie, Oeuvres*. Paris.
Rées, François Le. 1642. *Cursus philosophicus*. 3 vols. Paris.
Régis, Pierre-Sylvain. 1690. *Systême de philosophie, contenant la logique, la metaphisique, la physique et la morale*. 4 vols. Lyon.
——. 1691. *Réponse au livre qui a pour titre P. Danielis Huetii . . . Censura Philosophiae Cartesianae*. Paris.
——. 1970 [1691]. *Cours entier de philosophie; ou, Systeme general selon les principes de M. Descartes, contenant la logique, la metaphysique, la physique, et la morale*. Reprint. New York: Johnson.
Reif, Mary. 1962. *Natural Philosophy in Some Early Seventeenth Century Scholastic Textbooks*. Ph.D. Diss., Saint Louis University.
——. 1969. "The Textbook Tradition in Natural Philosophy, 1600–1650." *Journal of the History of Ideas* 30: 17–32.
Rimini, Gregory of. 1522. *Super primum et secundum sententiarum*. Venice.
Robert, Paul. 1953. *Dictionnaire alphabétique et analogique de la langue française*. Paris: Société du Nouveau Littré.
Rochemonteix, Camille de. 1889. *Un collège des Jesuites au XIIe et XIIIe siècle: Le collège Henri IV de la Flèche*. Le Mans.
Rodis-Lewis, Genevieve. 1985. *Idées et vérités éternelles chez Descartes*. Paris: Vrin.
——. 1987. "Descartes et les mathématiques au college." In *Le Discours et sa methode*, ed. N. Grimaldi and J.-L. Marion, 187–212. Paris: P.U.F.
Rohault, Jacques. 1987 [1723–24]. *Rohault's System of Natural Philosophy*. Reprint. New York: Garland.
Rorty, A. O., ed. 1986. *Essays on Descartes's Meditations*. Berkeley: University of California Press.
Ruler, J. A. van. 1995. *The Crisis of Causality: Voetius and Descartes on God, Nature, and Change*. Leiden: E. J. Brill.
Sarnanus, Constantinus. 1590. *Conciliatio dilvcida omnivm controversiarvm quae in doctrina duorum summorum theologorum S. Thomae & subtilis Ioannis Scoti passim leguntur*. . . . Lyon.
Scheiner, Christopher. 1626. *Rosa Ursina*. Bracciani.
Schmaltz, Tad. 1998. "Divine Causation and the Disappearance of Analogy: Descartes, Spinoza, Régis." Lecture delivered at the Southeastern Seminar in the History of Early Modern Philosophy, Blacksburg, VA, November 14–15, 1998.
Schmitt, Charles B. 1987. "Philoponus' Commentary on Aristotle's *Physics* in the Sixteenth Century." In *Philoponus and the Rejection of Aristotelian Science*, ed. R. Sorabji, 210–23. Ithaca, N.Y.: Cornell University Press.
Schmitt, Charles B., and Quentin Skinner, eds. 1987. *Cambridge History of Renaissance Philosophy*. Cambridge: Cambridge University Press.
Schofield, C. J. 1981. *Tychonic and Semi-Tychonic World Systems*. New York: Arno.
Scotus, John Duns. 1968 [1639]. *Opera Omnia*. Ed. L. Wadding. Reprint. Hildesheim: Georg Olms.
Sennert, Daniel. 1618. *Epitome naturalis scientiae*. Wittenberg.
——. 1659. *Thirteen Books of Natural Philosophy*. London.
Shapin, Steven. 1996. *The Scientific Revolution*. Chicago: University of Chicago Press.

Shapin, Steven, and Simon Schaffer. 1985. *Leviathan and the Air Pump: Hobbes, Boyle, and the Experimental Life*. Princeton: Princeton University Press.
Sirven, J. 1987. *Les années d'apprentissage de Descartes*. New York: Garland.
Smith's Smaller Latin-English Dictionary. 1968 [1855]. London: John Murray.
Sorel, Charles. 1634. *La science des choses corporelles, premiere partie de la science humaine ou l'on connoit la verité de toutes choses du monde par les forces de la raison et ou l'on treuve la refutation des erreurs de la philosophie vulgaire*. Paris.
Sorell, Tom. 1994. "Descartes's Modernity." In *Reason, Will, and Sensation: Studies in Descartes's Metaphysics*, ed. John Cottingham, 29–47. Oxford: Clarendon Press.
Sortais, Gaston. 1929. *Le cartésianisme chez les jésuites français au XVIIe et au XVIIIe siècle*. Paris: Beauchesne.
———. 1937. "Descartes et la Compagnie de Jésus, menaces et avances (1640–1646)." *Estudios de la Academia literaria del Platal Buenos Aires* 57: 441–68.
Sutton, Geoffrey V. 1995. *Science for a Polite Society: Gender, Culture, and the Demonstration of Enlightenment*. Boulder: Westview Press.
Swerdlow, N. 1993. "Science and Humanism in the Renaissance: Regiomontanus's Oration on the Dignity and Utility of the Mathematical Sciences." In *World Changes: Thomas Kuhn and the Nature of Science*, ed. P. Horwich, 131–68. Cambridge: MIT Press.
Thorndyke, Lynn. 1941–58. *A History of Magic and Experimental Science*. 8 vols. New York: Columbia University Press.
Toletus, Franciscus. 1572. *Commentaria una cum quaestionibus in universam Aristotelis logicam*. Venice.
———. 1574. *Commentaria una cum quaestionibus intres libros Aristotelis de amina*. Venice.
———. 1589. *Commentaria una cum quaestionibus in octo libros de Physica auscultatione*. Venice.
Urmson, J. O. 1967. "Ideas." *Encyclopedia of Philosophy*. Ed. Paul Edwards. New York: Macmillan.
Verbeek, Theo. 1992. *Descartes and the Dutch. Early Reactions to Cartesian Philosophy 1637–1650*. Carbondale: University of Southern Illinois Press.
———, ed. and trans. 1988. *La Querelle d'Utrecht*. Paris: Impressions Nouvelles.
Vigne, Guillelmus de la. 1666. Student thesis, Jesuit collège du Mont. Caen ms. 468.
Ville Louis de la [Louis le Valois]. 1680. *Sentimens de Monsieur Descartes touchant l'essence et es proprietez du corps opposez à la Doctrine de l'Eglise, et conforme aux erreurs de Calvin sur le sujet de l'Eucharistie*. Caen.
Vincent, Jean. 1660–71. *Cursus philosophicus, in quo totius scholae questiones fere omnes....* Toulouse.
———. 1677. *Discussio peripatetica in qua philosophiae cartesianae principia*. Toulouse.
Wallace, William. 1984. *Galileo and His Sources: The Heritage of the Collegio Romano in Galileo's Science*. Princeton: Princeton University Press.
Watson, Richard A. 1982. "Transubstantiation among the Cartesians." In *Problems of Cartesianism*, ed. T. Lennon, J. W. Nicholas, and J. W. Davis. Kingston: McGill-Queen's University Press.
———. 1987. *The Breakdown of Cartesian Metaphysics*. Atlantic Highlands, N.J.: Humanities Press.
Weisheipl, J. A. 1978. "The Nature, Scope, and Classification of the Sciences." In *Sciences in the Middle Ages*, ed. D. C. Lindberg, 461–82. Chicago: University of Chicago Press.

———. 1985. "Classification of the Sciences in Medieval Thought." In *Nature and Motion in the Middle Ages,* ed. W. E. Carroll, 203–37. Washington D.C.: Catholic University Press.

Wells, Norman. 1993. "Descartes' *Idea* and Its Sources." *American Catholic Philosophical Quarterly* 67: 513–35.

Westfall, Richard S. 1971. *The Construction of Modern Science: Mechanisms and Mechanics.* New York: Wiley.

Whitrow, G. J. 1988. *Time in History: The Evolution of Our General Awareness of Time and Temporal Perspective.* Oxford: Oxford University Press.

William of Ockham. 1967–86. *Opera Philosophica et Theologica.* St. Bonaventure, N.Y.: The Franciscan Institute.

William of Sherwood. 1968. *Treatise on Syncategorematic Words.* Trans. N. Kretzmann. Minneapolis: University of Minnesota Press.

Wilson, Margaret D. 1978. *Descartes.* London: Routledge.

Wolter, Allan B. 1990. *The Philosophical Theology of John Duns Scotus.* Ithaca, N.Y.: Cornell University Press.

Zavalloni, Roberto. 1951. *Richard de Media Villa et la controverse sur la pluralité des formes.* Louvain: Institut Supérieur de Philosophie.

Index

Abrégé de la philosophie de Gassendi (Bernier), 182–83, 186
Accidents, 49, 78, 81, 84–85, 89–91, 95, 129; condemnations and, 158–59, 163; transubstantiation and, 142, 150, 183–84
Acquaviva, Claudio, 17, 40
Actuality, 78–80, 129–30, 165–68
Albert of Saxony, 167, 168
Amy, L', Bernard, 177
André, Père, 34
Angels, 15, 84
Anti-Aristotelians, 87–88, 144–45, 163, 174
Aquinas, Thomas, 2; Aristotle's texts and, 178; in curriculum, 9; ideas and, 63–64; matter and, 80, 81, 83, 92; objective concept and, 43; place and, 49; on privation, 78; *ratio studiorum* and, 17; *Summa theologiae*, 13, 40, 150; void and, 132. *See also* Thomism
Aristotelianism, 1, 3–4, 34; anti-Aristotelians, 87–88, 144–45, 163, 174; astronomy and, 30, 100–103, 105–6, 108–10, 115–16; condemnations of, 156, 163, 178; corpuscularianism and, 124, 132; ideas and, 64; matter and, 87–88, 93–96; place and, 31–32n, 49–50; time and, 53–54
Aristotle: atomism and, 130–31; commentaries on, 26; in curriculum, 9, 12–13; infinity and, 165–67; *Physics*, 78, 145, 161–62; *Posterior Analytics*, 5–6, 191; on time, 54; transubstantiation and, 179; void and, 131–32, 161–62
Armogathe, J.-R., 57
Arnauld, Antoine, 4, 32, 40, 58, 173; on condemnations, 177–79; on Gassendi, 180; *Logique, ou l'art de penser, La*, 58; transubstantiation and, 140–41, 144, 147, 150, 179, 184
Astronomy, 3–4, 22; Aristotelianism and, 100–103, 105–6, 108–10, 115–16; eccentric-epicycle model, 101–3, 114; Jupiter, moons of, 20–21, 100, 103, 116–17; Mercury, 103–4, 106; moonspots, 101, 103, 106–9; novas and, 113; sublunary and superlunary worlds, 107, 110, 111; sunspots, 98–99, 103–9; telescope and, 100–101, 107, 109; Tychonic, 19n, 110, 114–15, 116–17; Venus, 101–3. *See also* Comets
Atomism, 14n, 21–22, 85–86, 157, 172; Aristotle and, 130–31; condemnation of, 174, 178–80. *See also* Basso, Sebastian; Corpuscularianism
Atoms, 133–34
Aubrey, John: *Brief Lives of Contemporaries*, 140
Augmentation and diminution, 79, 86, 131, 133
Augustine, 63, 72
Averroës, 49, 50, 52, 107

Babin, François, 175–76
Baillet, Adrien, 140–41, 153

[223]

Barlaeus, Caspar, 128
Barrant, Comte de, 10
Basso, Sebastian, 4, 85–86, 125–28, 133–36, 137; Mersenne on, 126–27; *Philosophiae naturalis adversus Aristotelem*, 126, 128, 139; rarefaction and, 127–28, 137–38
Bayle, Pierre, 181–82
Beck, L. J., 58
Being: infinity and, 167; intellect and, 46–48, 55–56; objective, 41, 44
Bellarmine, R., 100–101, 103
Bernier, François. *Abrégé de la philosophie de Gassendi*, 182–83, 186; *Doutes*, 182, 183; *Eclaircissement sur le livre de M. de la Ville*, 181–82, 184–85; transubstantiation and, 182–85
Bible, 13
Bitault, Jean, 87
Borgia, Francisco, 13–15
Bouillier, Francisque, 140, 182–83
Boujou, Théophraste, 50–51, 80, 107–8, 171n; *Corps de Philosophie*, 61–64, 111, 113, 146, 164
Bourdin, Pierre, 5, 24–26, 28–29, 156, 173; astronomy and, 114n; *cogito* and, 193–96, 203–4; in Seventh Set of Objections, 33, 193–96, 203–4
Brahe, Tycho de, 19n, 110, 112
Brief Lives of Contemporaries (Aubrey), 140
Brockliss, L. W. B., 168
Burgersdijk, Franco: *Institutionum metaphysicarum*, 85
Buridan, John, 130
Burman, Franz, 170, 196–97

Cajetan, Thomas, 41, 45, 56
Calvinists, 149, 153, 179
Carraud, Vincent, 92
Cartesians: condemnations and, 155–56, 187, 192, 203; doubt and, 156–57, 175, 180, 192; form-matter and, 93–96; political arena and, 175–78; satire, 176
Casimir of Toulouse, 125
Categorematic infinite, 167–69
Caterus, Johannes, 40
Catholic Church, authority of, 173–74
Causes, 14–16, 86, 134; ideas and, 61–66, 73–74
Ceriziers, René de, 54–55, 82–83, 106, 112, 170; *Le philosophe français*, 105, 145, 161–62

Charpentier, Nicolas, 126
Chouet, Jean-Robert: *Meletemata*, 94
Classification of sciences, 157, 191
Clave, Etienne de, 87, 88
Clavius, Christopher, 100–101, 107; *Sphaera*, 102
Clerselier, Claude, 141, 143, 147
Cogito: commensurate universal and, 5–6, 192–93; doubt and, 198–99, 203–5; as enthymeme, 192, 194, 197–98; God and, 197, 199, 200, 203; intellectual context, 188–91; knowledge and, 194–202; logic and, 197, 199; memory and, 199–200; as reflexive proposition, 200–202; *res cogitans*, 193–94; as simple intuition, 198, 200, 203; syllogism and, 195–97, 204; temporal nature of, 197–200
Coimbrans, 26, 27, 40, 146
Colegio das Artes, 26
Collectio judiciorum de novis erroribus (Duplessis d'Argentré), 163, 178
Collège d'Aix, 173
Collège de Clermont, 24–25, 33, 57, 142, 155–57, 191
Collège de Grassins, 27
Collège de Navarre, 27
Collège du Plessis, 27
College of Angers, 5, 142, 176–77
Collège Royal, 173
Collegio Romano, 100–101, 103
Comets: accidental changes and, 117; as fiery exhalations, 112–14; parallax measurement of, 110–12; tails, 112–13, 117–18; two theories of, 113–14
Commensurate universal, 5–6, 192–93
Commentary on the Physics (Toletus), 80
Common opinion, 14
Concept, 42–44, 65, 69–71
Conceptus, 65, 69
Condemnations, 5, 33–35, 144–45; accidents and, 158–59, 163; of Aristotelianism, 156, 163, 178; of atomism, 174, 178–80; Cartesians and, 155–56, 174, 187, 192, 203; extension and, 159–60; history of, 163–65, 174, 178; infinity and, 165–68; intellectual context, 155–56, 163–65, 172–73; by Louis XIV, 174, 175, 176, 189; political arena and, 156, 175–78; pragmatic reasons for, 156–57; substance and, 158–59, 165, 179, 180–81; transubstan-

INDEX [225]

tiation and, 142, 157, 163–64, 174–75, 179. *See also Index of Prohibited Books*
Conimbricenses, 128, 147
Copernicanism, 19, 100, 101
Corps de Philosophie (Boujou), 61–64, 111, 146, 164
Corpuscularianism, 4, 16, 77, 136, 172;
 Aristotelianism and, 124, 132;
 rarefaction and, 137. *See also* Atomism; Basso, Sebastian; Toletus, Franciscus
Correspondence (Descartes), 143–44
Corruption, 78, 79, 86, 129, 131, 133–34. *See also* Astronomy
Cosmology. *See* Astronomy
Council of Trent, 13, 143, 150, 151, 153, 183
Cousin, Victor, 177
Crassot, Jean, 3, 61; *Physics,* 70, 112
Cremonini, Cesare, 99, 100
Cursus philosophicus (Le Rées), 111–12
Cursus philosophicus (Vincent), 107

Dalbiez, Roland, 41–42, 45, 56
Daniel, Gabriel, 190
De Boot: *Philosophia Naturalis Reformata,* 89, 94
Democritus, 124–25
Derodon, David, 94
Descartes, René: on Basso, 127, 128, 138; on *cogito,* 193, 195, 196; on contemporaries, 34–35; on corpuscules, 136; on Democritus, 124; on doubt, 180; on form, 90–91, 92; on Galileo, 98–100; on idea, 58–61; infinity and, 170, 171; on Jesuits, 21, 25–26, 147–48; on knowledge, 97; on La Flèche, 7–10; matter and, 89–93; metaphysics of, 97–98, 138–39; on motion, 92, 136–37; on *novatores,* 85; request for objections, 10–12, 21, 24–26, 147–48; scholastic education, 7, 26–29, 40–41, 90; on Scholastics, 98–100; Scotism of, 55–57; on sunspots, 98–100; terminology and, 22–23, 29, 91–92; traditional concepts and, 90–91; on transubstantiation, 148–51; on void, 162;
—Works: *Correspondence,* 143–44; *Dioptrics,* 25, 149; *Essays,* 25, 148; *Fourth Objections,* 4, 149–51; *Fourth Set of Replies,* 40, 151; *Letter to Dinet,* 33, 147; *Meditation, Objections and Replies,* 193; *Meteors,* 89, 90; *Le Monde,* 30, 56, 90–92, 136; *Objections,* 4, 11, 29, 149–51, 193–96, 203–4; *Olympica,* 90; *Replies,* 11, 29, 32–33, 40, 90, 151, 193, 196; *Rules for the Direction of the Mind,* 29, 70; *Seventh Objections,* 33, 147–48, 193–96, 203–4. *See also Discourse on Method; Meditations on First Philosophy; Principles of Philosophy*
Dioptrics (Descartes), 25, 149
Discourse on Method (Descartes), 7, 10–11, 29–30, 147–48, 173; astronomy and, 30; *cogito* and, 205; form and, 91; objective being, 41
Disputationes metaphysicae (Suarez). *See Metaphysical Disputations*
Disquisitio Metaphysica (Gassendi), 194
Distinctions, theory of, 44–45
Doctrina ignorantia, De (Nicholas of Cusa), 170–71
Doubt, 156–57, 175, 180, 192; *cogito* and, 198–99, 203–5
Doutes sur quelques-uns des principaux chapitres de son abrégé de la philosophie de Gassendi (Bernier), 182, 183
Dualism, 139; of exemplars, 65–66, 70–71; matter and, 80–81, 83, 93
Du Chevreul, Jacques, 4, 101–4, 168
Duhamel, Jean, 5, 162; *Philosophia universalis,* 106–7, 118–19; *Quaedam recentiorum,* 163, 178; *Reflexions critiques sur le système cartésien,* 200–204
Du Moulin, Pierre, 112–13
Dupleix, Scipion, 27n, 41n, 52, 81–82, 128, 152, 170; astronomy and, 110; *Métaphysique,* 83–84; *Physique,* 83, 107, 145, 164; place and, 145–46
Duplessis d'Argentré, Charles: *Collectio judiciorum de novis erroribus,* 163, 178
Durandus a Sancto Porciano, 42, 43

Eclaircissement sur le livre de M. de la Ville (Bernier), 181–82, 184–85
Eidos, 66, 68–70
Elements, 14, 16, 19, 88, 91, 95; atoms, 133–34; rarefaction and, 130–31. *See also* Ether
Entire Body of Philosophy according to the Principles of the Famous Renate Descartes (Le Grand), 95–96
Essays (Descartes), 25, 148
Ether, 4, 24n, 86, 127–28, 134–36, 138. *See also* Elements
Eucharist. *See* Transubstantiation

Eustachius a Sancto Paulo, 2–3, 26–29, 43–44, 128; astronomy of, 102–3; ideas and, 61, 64–65; on infinity, 168–69; matter and, 81, 84–85; *Philosophy*, 27; *Physics*, 64; Scotism and, 47–48, 51, 57; *Summa philosophiae quadripartita*, 27, 39–41, 64–65, 102–3, 201n; transubstantiation and, 146, 148, 152
Exemplars, 62–66; archetype and, 68, 73–74; dianoetic, 70–71; duality of, 65–66, 70–71; image and, 66–69; *ratio* and, 66–67, 72; species and, 65–70; truth and, 65–66, 74
Exercitationes paradoxicae adversus Aristoteleos (Gassendi), 164, 178
Extension, 5, 139, 159–60, 185; Gassendists and, 181, 183–84; substance and, 91, 93, 95–96

Fabri, Honoré, 125, 179
Faith, 17–18, 30, 179
Fonseca, Petrus, 66n
Form, 3, 16; actuality and, 78–80, 129–30; Aristotelianism and, 87–89, 93–94; Cartesians and, 93–96; condemnations and, 158–59, 179; creation and, 82–83; Descartes on, 90–92; hylomorphism and, 3, 77; ideas and, 61–63, 71–75; individuation and, 3, 85; plurality of, 55–56; substantial forms, 78, 88, 129, 139, 157. *See also* Ideas; Matter; Place; Substance
Formal concept, 42–44, 65, 69, 71–75
Formarum Origine, De (Pemble), 89
Fourth Objections (Descartes), 4, 149–51
Fourth Set of Replies (Descartes), 40, 151
Frey, J.-C., 87, 88
Fromondus, Libertius, 21–22, 24

Galileo Galilei, 21, 30, 97–99, 119, 139; comets and, 110–11, 112; sunspots and, 99
Gassendi, Pierre, 5, 146–47, 172–73; on *cogito*, 193; *Disquisitio Metaphysica*, 194; *Exercitationes paradoxicae adversus Aristoteleos*, 164, 178
Gassendists, 5, 172; extension and, 181, 183–84; place and, 182, 185–86; transubstantiation and, 182–85
Gaultruche, Pierre, 52–53, 83
Generation and corruption, 78–79, 86, 129, 131, 133–34. *See also* Astronomy

Gilson, Etienne: *Index scolastico-cartésien*, 2, 39, 56–57; critique of, 41, 45
Goclenius, Rudolph: *Lexicon*, 3, 65–69
God: being and, 46–48, 138; *cogito* and, 197, 199, 200, 203; conception and, 69–70; ether and, 135–36; ideas and, 3, 59, 61–64, 67, 71–74; infinity and, 166–70; matter and, 80–83, 86, 89, 145; motion and, 137, 139; plurality of worlds and, 160, 171; in *ratio studiorum*, 14, 16; void and, 131–32, 161–62, 163
Gorlaeus, David, 126
Goudin, Antoine, 83, 105–6, 106–9, 115, *Philosophie*, 105–6
Grandamy, Jacques, 115–18
Grange, Jean-Baptiste de la, 102, 142–43; *Les principes de la philosophie*, 189–90
Grassi, Horatio, 110, 112
Gregory of Rimini, 167, 168
Grene, Marjorie, 3

Haecceitas, 84
Hamlyn, Octave, 75
Harlay, François de, 174
Heaven, 49–53, 107–8, 114–15
Heereboord, Adriaan, 89, 94
Henry IV, 19–21, 100
Hill, Nicholas, 126
Hobbes, Thomas, 5, 58, 63, 140, 193–94
Homocentric spheres, 18–19
Huet, Pierre Daniel, 5, 197–98, 204, 205
Huygens, Constantijn, 128; correspondence, 21–22, 25, 128, 148–50
Hylomorphism, 3, 77, 83, 89–90

Ideas: causes and, 61–66, 73–74; definitions, 58–61; Descartes on, 58–61; *eidos* and, 66, 68–70; as exemplars, 62–66; form and, 61–63, 71, 75; God and, 3, 59, 61–64, 67, 71–74; image and, 65–69, 74; literary terminology, 74, 76; material falsity and, 32–33; objective *vs.* formal concept, 42–44, 65, 69, 71–74; Platonism and, 58–59, 62, 64, 66, 71; senses and, 68, 75–76; soul and, 63–64; truth and, 65–66, 74–75. *See also* Form
Impiété des Deistes, L' (Mersenne), 126
Impetus, 130, 134, 138
Index of Prohibited Books, 5, 34, 142, 155, 157, 173, 180. *See also* Condemnations

INDEX   [227]

Index scolastico-cartésien (Gilson), 2, 39, 41, 56–57
Individuation, 3, 77–78, 83–85, 152–53
Infinity, 165–71
Inquisition, 187
Institutionum metaphysicarum (Burgersdijk), 85
Intellectual context, 1–2, 6, 48; *cogito* and, 188–91; condemnations and, 155–56, 163–65, 172–73; Gassendists and, 186–87; transubstantiation and, 140–44, 151–54

Jansenists, 177. *See also* Arnauld, Antoine
Jesuit pedagogy, 1, 12–21; condemnations and, 159–60; faith and, 17–18, 30, 179; novelties and, 12, 15–20; progressive doctrines, 18–19; *ratio studiorum* and, 12–17, 40; textbooks, 12–13, 26–27; Thomistic philosophy and, 13, 17, 39, 40, 48; transubstantiation and, 142–45, 159. *See also* La Flèche; University of Paris
Jolley, Nicholas, 141
Judgment, 74–75
Jupiter, 20–21, 100, 103, 116

Knowledge, *cogito* and, 194–202

Lacrymae Collegii Flexiensis, 20–21, 100
La Flèche, 4, 33–34, 57, 100, 114, 173; curriculum, 8–9, 12–13; equality and, 9–10; Thomist teachings and, 13, 17, 39–40, 48; transubstantiation and, 147. *See also* Jesuit pedagogy
Launoy, Jean de: *De varia Aristotelis fortuna*, 163, 178
Le Bossu, René, 93
Le Grand, Antoine: *Entire Body of Philosophy according to the Principles of the Famous Renate Descartes*, 95–96
Leibniz, G. W., 201n, 205
Leo XIII, 57
Le Rées, François: *Cursus philosophicus*, 111–12
Lesclache, Louis de, 112
Letter to Dinet (Descartes), 33, 147
Lexicon (Goclenius), 65–69
Locke, John, 58
Logic: *cogito* and, 197, 199; terminology of, 166
Logique, ou l'art de penser, La (Arnauld and Nicole), 58
Louis XI, 178

Louis XIV, 174, 175, 189
Loyola, Ignatius of, 13, 17, 40

Magnus, Albertus, 178
Maignan, Emanuel, 179
Material falsity, 32–33, 40
Mathematics, 22, 25n, 191
Matter, 3, 16; accidental, 78, 81, 84–91, 95, 129, 142; anti-Aristotelianism and, 87–89; Aristotelianism and, 87–88, 93–96; Cartesians and, 93–96; creation and, 82–83; Descartes's view, 89–93; dualism and, 80–81, 83, 93; extension and, 5, 91, 93, 95–96, 159–60, 183, 185; generation and, 78–79; God and, 80–83, 86, 89, 145; hylomorphism and, 3, 77, 83, 89–90; individuation and, 3, 77–78, 83–84; potentiality and, 79–81; prime matter, 144–45; quantity and, 78–79, 84, 86, 91, 129, 131; Scotism and, 80, 81, 83; solidity of, 183–86; substance and, 5, 78, 80, 88–91; subtle matter, 24; Thomism and, 81–84, 92, 145. *See also* Form; Motion
Mechanical philosophy, 16; historical context, 123–24. *See also* Atomism; Corpuscularianism
Meditation, Objections and Replies (Descartes), 193
Meditations on First Philosophy (Descartes), 11, 28, 29; *cogito* and, 193, 196, 205; condemnations and, 159; ideas and, 60, 69–70, 73, 74; material falsity, 32; matter and, 89; transubstantiation and, 147, 149–50
Meletemata (Chouet), 94
Mente concepta, 69
Mercury, 103–4, 106
Mersenne, Marin, 23–25, 87–88, 164, 172; on Aristotle, 126–27; correspondence, 4, 26–28, 39–40, 60, 71, 97, 138; *L'impiété des Deistes*, 126; transubstantiation and, 149–50; *La Vérité des sciences*, 126, 163
Mesland, Denis, 141–44, 154
Metaphysical Disputations (Suarez), 32–33, 39–43
Metaphysics: astronomy and, 97–98; being and, 47; ideas and, 75–76; theology and, 189–91
Métaphysique (Dupleix), 83–84
Meteors (Descartes), 89–90

Minima naturalia, 130–35
Monde, Le (Descartes), 30, 56, 90–92, 136
Moonspots, 101, 103, 106–9
Morin, Jean Baptist, 21–24, 26, 87–88
Motion, 30–31, 78–79; actuality and, 129–30; Descartes on, 92, 136–37; ether and, 86; God and, 137, 139; potentiality and, 129; rest and, 136–37; Scotist view, 48–53; universe and, 52–53; void and, 137–38, 161–62. *See also* Matter; Place; Space; Time
Mutation, 129. *See also* Corruption; Generation and corruption

Natural agents, 14, 16, 62
Nature, 82, 91
Nicholas of Cusa: *De doctrina ignorantia*, 170–71
Nicole, Pierre, 58
Ninth Congregation, 33
Noël, Etienne, 11n
Normore, Calvin, 55
Novas, 113
Novatores, 4, 85, 127–28
Novelties, 12, 15–20

Objections, 4, 11, 29, 149–51, 157, 193–96, 196
Objective being, 41, 44
Objective concept, 42–44, 65, 69, 71–74
Ockham, William of, 14n, 15n, 168
Olympica (Descartes), 90
Oratorian Fathers, 142, 155, 159–60, 192
Ovid, 82

Parlement, 144, 163
Patrizi, Franciscus, 126
Pélaut, 177
Pemble, William: *De Formarum Origine*, 89, 94
Philoponus, 52
Philosophe français, Le (de Ceriziers), 105, 145, 161–62
Philosophiae naturalis adversus Aristotelem (Basso), 126, 128, 139
Philosophia Naturalis Reformata (de Boot brothers), 89, 94
Philosophia universalis (Duhamel), 106–7, 118–19
Philosophie (Goudin), 105–6
Philosophy (Eustachius a Sancto Paulo), 27
Physics (Aristotle), 78, 145, 161–62
Physics (Crassot), 70, 112

Physics (Eustachius a Sancto Paulo), 64
Physics (Toletus), 168
Physique (Dupleix), 83, 145, 164
Piccolomini, Francesco, 33
Pius V, 13, 40–41, 57
Place, 31; Aristotelianism and, 31–32n, 49, 50; Gassendist view, 182, 185–86; internal *vs.* external, 50–52, 78–79, 129; *lieu de situation*, 50–51; Thomism and, 49–53, 164. *See also* Space; Transubstantiation
Plato: *Timaeus*, 49, 73, 75
Platonism, ideas and, 58–59, 62, 64, 66, 71
Plempius, 21, 22
Plurality of worlds, 160, 165, 171, 190
Port Royal Logic. *See Logique, ou l'art de penser, La*
Posterior Analytics (Aristotle), 5–6, 191
Potentiality, 78–81, 129, 165–69
Potier, Charles, 24
Pourchot, Edmond, 155
Les principes de la philosophie contre les nouveaux philosophes (de la Grange), 189–90
Principles of Philosophy (Descartes), 4, 28–29, 32, 56, 76, 155; astronomy and, 30, 98; *cogito* and, 196; condemnations and, 159–61; corpuscularianism and, 124; form and, 91–92, 96; place and, 31; transubstantiation and, 152, 154
Privation, 78, 93

Quaedam recentiorum philosophorum (Duhamel), 163, 178
Quaestiones, 13
Quality, 78–79
Quantity, 78–79, 84, 86, 91, 129, 131; transubstantiation and, 146, 164

Rabelais, François, 59
Raconis, Charles François d'Abra de, 2–3, 51; ideas and, 61, 71–74; matter and, 81, 84–85; philosophy of, 27, 42–45; *Summa totius philosophiae*, 27, 42–43; transubstantiation and, 146, 147, 152
Raey, Johannes de, 94, 95
Rapin, René, 156
Rarefaction, 4, 127–28, 130–38
Ratio, 66–67, 72
Ratio studiorum, 12–17, 26, 40
Reflexions critiques sur le système cartésien de la philosophie de mr. Régis (Duhamel), 200–204

Regiomontanus, 110
Régis, Pierre-Sylvan, 95n, 96, 198–203
Regius, Henricus, 90, 139, 157
Replies (Descartes), 11, 29, 32–33, 40, 90, 151, 158, 193, 196; transubstantiation and, 141, 147–48, 153
Res cogitans, 193–94
Rochemonteix, Camille de, 154
Rohault, Jacques, 93; *Traité de la physique,* 94–95
Rubius, Antonius, 26, 39
Rules for the Direction of the Mind (Descartes), 29, 70

Scholastics: form-matter dispute and, 79–80; Galileo and, 98–99. *See also* Aristotelianism; Aristotle; Jesuit pedagogy; La Flèche; Scotism; Thomism; University of Paris
Sciences, classification of, 157, 191
Scientific revolution, 97
Scotism, 2–3, 41, 44–47; of Descartes, 55–57; matter and, 80, 81, 83; motion and, 48–53; time and, 53–55; transubstantiation and, 152–54
Scotus, John Duns. *See* Scotism
Seneca, 67, 70
Sennert, Daniel, 128
Seventh Objections (Descartes, Bourdin), 33, 147–48, 193–96, 203–4
"Several Reasons for Preventing the Censure or Condemnation of Descartes' Philosophy," 177–79
Sixth Objections (Descartes), 5, 193–94
Skeptics, 204
Souls, 15–16, 48n, 151, 175; ideas and, 63–64; matter and, 80, 85
Space, 91, 190; Aristotelianism and, 31–32n, 49–50; Gassendist view, 182, 185–86; Scotist view, 48–53. *See also* Place
Species, 65–70, 91, 183–84
Sphaera (Clavius), 102
Suarez, Francisco: *Metaphysical Disputations,* 32–33, 39–43
Subjective concept, 42–44
Substance, 5, 80; condemnations and, 158–59, 165, 179–81; extension and, 91, 93, 95–96; substantial forms, 78, 88–91, 129, 139, 157; transubstantiation and, 142, 150, 158–59
Subtle matter, 24

Summa philosophica quadripartita (Eustachius a Sancto Paulo), 27, 39–41, 64–65, 102–3, 201n
Summa theologiae (Aquinas), 13, 40, 150
Summa totius philosophaie (de Raconis), 27, 42–43
Sunspots, 98–99, 103–9
Syncategorematic infinite, 167–69

Tamburini, Michel-Angelo, 34
Tarin, Jean, 10
Telescope, 100–101, 107, 109
Terminology, 22–23, 29, 74, 76, 166
Textbooks, 12–13, 26–27, 40–41, 64; astronomy and, 102–9; comets and, 111; French-language, 41n; plurality of worlds and, 160; transubstantiation and, 146, 152; void and, 161–62
Theology, metaphysics and, 189–91
Thomism, 2, 45–46; ideas and, 63; infinity and, 167–68; Jesuit pedagogy and, 13, 17, 39–40, 48; material falsity and, 32; matter and, 81–84, 92, 145; place and, 49–53, 164; time and, 54–55; transubstantiation and, 150–52. *See also* Aquinas, Thomas
Timaeus (Plato), 49, 73, 75
Time, 53–56, 197–200
Toletus, Franciscus, 26, 39, 50, 54, 128, 130, 132, 138; *Commentary on the Physics,* 80; *Physics,* 168
Tournemine, Jean, 18, 114
Traité de la physique (Rohault), 94–95
Transubstantiation, 4–5; accidents and, 142, 150, 183–84; Christ in Eucharist, 150–51; condemnations and, 142, 157, 163–64, 174–75, 179; Descartes on, 148–51; Gassendists and, 182–85; individuation and, 152–53; intellectual context, 140–44, 151–54; Jesuit pedagogy and, 142–45, 159; place and, 145–46; quantity and, 146, 164; Scotism and, 152–54; substance and, 142, 150, 158–59
Tremblay, Ignace de, 34
Trésor de la langue française tant ancienne que moderne (Nicot), 59
Truth: exemplars and, 65–66, 74; ideas and, 65–66, 74–75; material falsity and, 32; objective intellect and, 42

Ubi, 50–52, 78, 129. *See also* Place
Ultimate sphere (heaven), 49–53, 107, 114–15
Universality, 84, 192
Universe, 5, 18–19; infinity and, 170–71; motion and, 52–53; plurality of worlds, 160, 165, 171, 190. *See also* Astronomy
University of Frankener, 56
University of Louvain, 5, 142–43, 155, 157–60
University of Paris (Sorbonne), 13, 45, 48, 56–57, 64, 87; condemnations, 159–60, 163, 175, 177, 180; transubstantiation and, 144–45
Utrecht University, 32, 90, 155, 156, 157

Valois, Père de (Louis de la Ville): *Sentimens de Monsieur Descartes*, 143
Varia Aristotelis fortuna, De (de Launoy), 163, 178
Vasquez, Gabriel, 42
Velitatio, 24
Venus, 101–3
Vérité des sciences, La (Mersenne), 126, 163
Ville, Louis de la, 181–85; *Sentimens de Monsieur Descartes*, 143
Villon, Antoine, 87
Vincent, Justus: *Cursus philosophicus*, 107
Voetius, 32
Void, 4, 50, 131–32, 134, 146, 186; motion and, 137–38, 161–63

Watson, Richard, 141–42, 154